I0488087

NUREG-1835
Supplement 1

Safety Evaluation Report for an Early Site Permit (ESP) at the North Anna ESP Site

Manuscript Completed: November 2006
Date Published: November 2006

Division of New Reactor Licensing
Office of New Reactors
U.S. Nuclear Regulatory Commission
Washington, DC 20555-0001

ABSTRACT

The final safety evaluation report (FSER) (NUREG-1835, "Safety Evaluation Report for an Early Site Permit (ESP) at the North Anna ESP Site," issued September 2005) documents the U.S. Nuclear Regulatory Commission (NRC) staff's technical review of the site safety analysis report and emergency planning information included in the early site permit (ESP) application submitted by Dominion Nuclear North Anna, LLC (Dominion or the applicant), for the North Anna ESP site. By letter dated September 25, 2003, Dominion submitted the ESP application for the North Anna site in accordance with Subpart A, "Early Site Permits," of Title 10, Part 52, "Early Site Permits; Standard Design Certifications; and Combined Licenses for Nuclear Power Plants," of the *Code of Federal Regulations* (10 CFR Part 52). The North Anna ESP site is located approximately 40 miles north-northwest of Richmond, Virginia, and is adjacent to two existing nuclear power reactors operated by Virginia Electric and Power Company, which, like Dominion Nuclear North Anna, LLC, is a subsidiary of Dominion Resources, Inc. In its application, Dominion seeks an ESP that could support a future application to construct and operate one or more additional nuclear power reactors at the ESP site.

The NRC staff has prepared this supplement to the FSER because Dominion amended its ESP application in Revisions 6, 7, 8, and 9. In Revision 6, which the applicant submitted to the NRC on April 13, 2006, Dominion described a new approach for cooling its proposed Unit 3. Under the revised approach, Unit 3 would use a closed-cycle cooling system, rather than the originally proposed once-through cooling system. Dominion also planned to increase the power level of both proposed Units 3 and 4 from 4300 megawatts thermal (MWt) to 4500 MWt (the designed maximum power of a General Electric Economic and Simple Boiling-Water Reactor (ESBWR)) with a total nuclear generating capacity of up to 9000 MWt.

This supplemental FSER includes the NRC staff's analysis of the safety aspects of constructing and operating a closed-cycle cooling system for Unit 3 and the increase in power for proposed Units 3 and 4. This supplement presents the results of the staff's review of information submitted in conjunction with the ESP application. The staff has identified, in Appendix A to this supplement, certain site-related items that an applicant will need to address at the combined license or construction permit stage, should it desire to construct one or more new nuclear reactors on the North Anna ESP site. The staff determined that these items do not affect the staff's regulatory findings at the ESP stage and are more appropriately addressed at later stages in the licensing process. In addition, Appendix A to this supplement also identifies the proposed permit conditions that the staff recommends the Commission impose, should the NRC issue an ESP to the applicant.

In this supplement, the staff has evaluated only the safety aspects of the changes in cooling design and the increase in power level presented in Revisions 6, 7, 8, and 9 of the ESP application. This supplement contains only those sections and/or chapters of the FSER that are affected by the changes presented in Revisions 6, 7, 8 and 9 of the ESP application. The NRC revised Appendix A and provides it in its entirety for clarity and ease of inclusion in any ESP that the NRC may issue.

CONTENTS

Appendices

Tables

EXECUTIVE SUMMARY

The staff has revised this executive summary to reflect the changes Dominion Nuclear North Anna, LLC, presented in Revisions 6, 7, 8, and 9 of its early site permit application for proposed North Anna Units 3 and 4.

On September 25, 2003, the U.S. Nuclear Regulatory Commission (NRC) received an application from Dominion Nuclear North Anna, LLC (Dominion or the applicant), for an early site permit (ESP) for two units located adjacent to the North Anna Power Station, Units 1 and 2. The North Anna ESP site is located approximately 40 miles north-northwest of Richmond, Virginia, and is adjacent to two existing nuclear power reactors operated by Virginia Electric and Power Company, which, like Dominion Nuclear North Anna, LLC, is a subsidiary of Dominion Resources, Inc.

Title 10, Part 52, "Early Site Permits; Standard Design Certifications; and Combined Licenses for Nuclear Power Plants," of the *Code of Federal Regulations* (10 CFR Part 52) contains requirements for licensing, construction, and operation of new nuclear power plants.[1] These regulations address ESPs, design certifications, and combined licenses (COLs). The ESP process (Subpart A, "Early Site Permits," of 10 CFR Part 52) is intended to address and resolve site-related issues. The design certification process (Subpart B, "Standard Design Certifications," of 10 CFR Part 52) provides a means for a vendor to obtain NRC certification of a particular reactor design. Finally, the COL process (Subpart C, "Combined Licenses," of 10 CFR Part 52) allows an applicant to seek authorization to construct and operate a new nuclear power plant. A COL may reference an ESP, a certified design, both, or neither. It is incumbent on a COL applicant to resolve issues related to licensing that were not settled as part of an ESP or design certification proceeding before the NRC can issue a COL.

The safety evaluation report (SER) (NUREG-1835, "Safety Evaluation Report for an Early Site Permit (ESP) at the North Anna ESP Site," issued September 2005) describes the results of a review by NRC staff based on Revision 5 of the ESP application submitted by Dominion for the North Anna site. The staff's review verified the applicant's compliance with the requirements of Subpart A to 10 CFR Part 52.

The NRC regulations also contain requirements for an applicant to submit an environmental report pursuant to 10 CFR Part 51, "Environmental Protection Regulations for Domestic Licensing and Related Regulatory Activities." The NRC reviews the environmental report as part of the Agency's responsibilities under the National Environmental Policy Act of 1969, as amended. The NRC presents the results of that review in a final environmental impact statement, which is a report separate from the final safety evaluation report (FSER).

In accordance with 10 CFR Part 52, Dominion submitted information in its ESP application that includes (1) a description of the site and nearby areas that could affect or be affected by a nuclear power plant or plants located at the site, (2) a safety assessment of the site on which the facility would be located, including an analysis and evaluation of the major structures, systems,

---

[1] Applicants may also choose to seek a construction permit and operating license in accordance with 10 CFR Part 50, "Domestic Licensing of Production and Utilization Facilities," instead of using the 10 CFR Part 52 process.

and components of the facility that bear significantly on the acceptability of the site, and (3) the proposed major features of emergency plans. The application describes how the site complies with the requirements of 10 CFR Part 52 and the siting criteria of 10 CFR Part 100, "Reactor Site Criteria."[2]

On April 13, 2006, Dominion submitted Revision 6 to its application, which included a revised site safety analysis report (SSAR) and environmental report. In Revision 6, Dominion proposed (1) changing its cooling design for proposed Unit 3 from a once-through cooling system, as described in previous versions of the SSAR, to a closed-cycle system and (2) increasing the maximum power output per unit from 4300 megawatts thermal (MWt) to 4500 MWt for proposed Units 3 and 4 (hereafter referred to as Units 3 and 4). Under the revised cooling system approach, Unit 3 would use a closed-cycle, combination wet and dry cooling system. Unit 4 will use the dry cooling system, as originally proposed. The proposed increase in power level corresponds to the revision of the designed maximum power (4,500 MWt) of a General Electric Economic and Simple Boiling-Water Reactor (ESBWR), one of the reactor designs included in the plant parameter envelope and evaluated in the FSER. The NRC staff decided to prepare a supplement to the FSER to evaluate the impact of the above changes. The staff, in its review of Revision 6 of the application, requested additional information from Dominion. Dominion responded to the requests for additional information (RAIs) and on June 21 and July 31, 2006, submitted Revisions 7 and 8, respectively, to the application, which included the necessary information from the RAI responses.

This supplemental FSER includes the NRC staff's analysis of the safety aspects of constructing and operating a closed-cycle cooling system for Unit 3 and the increase in power for Units 3 and 4. This supplement presents the results of the staff's review of information submitted in conjunction with the ESP application. The staff has identified, in Appendix A to this supplement, certain site-related items that an applicant will need to address at the combined license or construction permit stage, should it desire to construct one or more new nuclear reactors on the North Anna ESP site. The staff determined that these items do not affect the staff's regulatory findings at the ESP stage and are more appropriately addressed at later stages in the licensing process. In addition, Appendix A to this supplement identifies the proposed permit conditions that the staff recommends the Commission impose, should the NRC issue an ESP to the applicant.

In this supplement, the staff has evaluated only the safety aspects of the change in cooling design and the increase in power level presented in Revisions 6, 7, 8, and 9 of the ESP application. This supplement contains only those sections and/or chapters of the FSER that are affected by the changes presented in Revisions 6, 7, 8 and 9 of the ESP application. The staff has revised Appendix A and provides it in its entirety for clarity and ease of inclusion in any ESP that the NRC may issue.

---

[2] The applicant has also submitted information intended to partially address some of the general design criteria (GDC) in Appendix A, "General Design Criteria for Nuclear Power Plants," to 10 CFR Part 50, "Domestic Licensing of Production and Utilization Facilities." Only GDC 2, "Design Bases for Protection Against Natural Phenomena," applies to an ESP application, and it does so only to the extent necessary to determine the safe-shutdown earthquake (SSE) and the seismically induced flood. The staff has explicitly addressed partial compliance with GDC 2, in accordance with 10 CFR 52.17(a)(1) and 10 CFR 50.34(a)(12), only in connection with the applicant's analysis of the SSE and the seismically induced flood. Otherwise, an ESP applicant need not demonstrate compliance with the GDC. The staff has included a statement to this effect in those sections of the FSER that do not relate to the SSE or the seismically induced flood. Nonetheless, this FSER describes the staff's evaluation of information submitted by the applicant to address GDC 2.

The NRC's Advisory Committee on Reactor Safeguards (ACRS) reviewed the bases for the conclusions in this supplemental report. The ACRS independently reviewed those aspects of the application that concern safety, as well as this supplement to the safety evaluation report, and provided the results of its review in the report dated October 13, 2006. Appendix E includes a copy of the report by the ACRS on the final safety evaluation, as required by 10 CFR 52.53, "Referral to the ACRS."

ABBREVIATIONS

ABWR	advanced boiling-water reactor
a	acre
ACR-700	Atomic Energy of Canada Advanced CANDU Reactor
ADAMS	Agencywide Documents Access and Management System
ALARA	as low as is reasonably achievable
ALI	annual limits on intake
ALWR	advanced light-water reactor
ANS	alert and notification system
ANSI	American National Standards Institute
ANSS	Advanced National Seismic System
AP1000	Westinghouse Advanced Plant 1000
ARA	Applied Research Associates
ASCE	American Society of Civil Engineers
ASHRAE	American Society of Heating, Refrigerating and Air-Conditioning Engineers
ASME	American Society of Mechanical Engineers
ASTM	American Society for Testing and Materials
BRH	Bureau of Radiological Health
BWR	boiling-water reactor
CCW	component cooling water
CDE	committed dose equivalent
CEDE	committed effective dose equivalent
CEUS	central and eastern United States
CFR	*Code of Federal Regulations*
CFS	cubic feet per second
CLB	cleanup waterline break
COL	combined license
COVRERP	Commonwealth of Virginia Radiological Emergency Response Plan
CP	construction permit
CPT	cone penetrometer test
CVSZ	Central Virginia seismic zone
DAC	derived air concentration
DBA	design-basis accident
DCD	design control document
DEIS	draft environmental impact statement
DEM	Department of Emergency Management
DOE	Department of Energy
DSER	draft safety evaluation report
EAB	exclusion area boundary
EAC	evacuation assembly center
EAL	emergency action level
EAS	emergency alert system
ECFS	East Coast fault system
EDE	effective dose equivalent
EDP	engineering department procedure
EDPI	engineering department instructions
EDS	engineering design spectrum
EIS	environmental impact statement

EMI	Emergency Management Institute
ENS	emergency notification system
EOC	emergency operations center
EOF	emergency operations facility
EPA	Environmental Protection Agency
EPDS	electronic procedure distribution system
EPIP	emergency plan implementing procedure
EPRI	Electric Power Research Institute
EPZ	emergency planning zone
ER	environmental report
ERDS	Emergency Response Data System
ERO	emergency response organization
ESBWR	Economic and Simple Boiling-Water Reactor
ESE	east-southeast
ESIM	evacuation simulation model
ESP	early site permit
EST	earth science team
ETE	evacuation time estimate
ETSZ	Eastern Tennessee seismic zone
FAA	Federal Aviation Administration
FEMA	Federal Emergency Management Agency
FLB	feedwater line break
FPS	feet per second
FRERP	Federal Radiological Emergency Response Plan
FRMAC	Federal Radiological Monitoring and Assessment Center
FRP	Federal Response Plan
FS	factor of safety
FSER	final safety evaluation report
ft	foot/feet
gal	gallon
GBU	Global Business Unit
GDC	general design criterion
GE	General Electric
GIS	geographic information system
GPM	gallons per minute
GSA	Geological Society of America
GT-MHR	Gas Turbine Modular Helium Reactor
HEAR	hospital emergency and administrative radio
HEC	Hydrologic Engineering Center
HMR	hydrometeorological report
HPN	health physics network
Hz	hertz
IEM	Innovative Emergency Management, Inc.
in.	inch
in./mo	inch per month
INPO	Institute of Nuclear Power Operations
IRIS	International Reactor Innovative and Secure Reactor
ISFSI	independent spent fuel storage installation
ISO	International Organization for Standardization
KI	potassium iodide

km	kilometer
kPa	kilopascals
ksf	kip(s) per square foot
lb	pound
lbf/ft	pound-force per square foot
LFA	lead Federal agency
LLNL	Lawrence Livermore National Laboratory
LOCA	loss-of-coolant accident
LPZ	low-population zone
LWR	light-water reactor
m	meter
M&TE	measuring and test equipment
MCVH	Medical College of Virginia Hospitals
MEI	maximally exposed individual
mi/h	miles per hour
MIDAS	meteorological information and dose assessment system
MMI	modified mercalli intensity
mrem	millirem
MSL	mean sea level
mSv	millisievert
MWt	megawatt thermal
NAEP	North Anna Emergency Plan
NAPS	North Anna Power Station
NBU	Nuclear Business Unit
NCDC	National Climatic Data Center
NDCM	Nuclear Design Control Manual
NDCP	Nuclear Design Control Program
NGVD	National Geodetic Vertical Datum
NE	northeast
NEI	Nuclear Energy Institute
NEP	nuclear emergency preparedness
nmi	nautical mile
NMSZ	New Madrid seismic zone
NNE	north-northeast
NOAA	National Oceanic and Atmospheric Administration
NPSEPT	Nuclear Power Station Emergency Preparedness Training
NQAM	Nuclear Quality Assurance Manual
NRC	U.S. Nuclear Regulatory Commission
NRRL	nuclear-required records list
NSSL	National Severe Storms Laboratory
NUPIC	Nuclear Utility Procurement Issues Committee
NWS	National Weather Service
OBE	operating-basis earthquake
ODEC	Old Dominion Electric Cooperative
OL	operating license
OREMS	Oak Ridge Evaluation Modeling System
OSC	operational support center
OW	observation well
PAG	protective action guideline
PAR	protective action recommendation

PAZ	protective action zone
PBMR	pebble bed modular reactor
PGA	peak ground acceleration
PMCL	protective measures counterpart link
PMF	probable maximum flood
PMH	probable maximum hurricane
PMP	probable maximum precipitation
PMWP	probable maximum winter precipitation
PNNL	Pacific Northwest National Laboratories
PO	purchase order
PPE	plant parameter envelope
PPR	potential problem reporting
PQAM	project quality assurance manager
PSHA	probabilistic seismic hazard analysis
psi	pound per square inch
PWR	pressurized-water reactor
QA	quality assurance
QAPD	quality assurance program description
QAPP	quality assurance program plan
RAA	remote assembly area
RAI	request for additional information
RAP	radiological assistance program
REI	Risk Engineering, Inc.
RERP	radiological emergency response plan
RERT	Radiological Emergency Response Team
RIC	Richmond International Airport
RG	regulatory guide
RQD	rock quality designation
RS	review standard
RSCL	reactor safety counterpart link
s	second
S	south
SCC	State Corporation Commission
SCR	stable continental regions
SCS	Soil Conservation Service
SE	southeast
SEI	Structural Engineering Institute
SER	safety evaluation report
SF	scale factor
SPT	standard penetration test
SQAP	software quality assurance plan
SRCC	Southern Regional Climate Center
SSAR	site safety analysis report
SSC	structure, system, and component
SSE	safe-shutdown earthquake
SSHAC	Senior Seismic Hazard Advisory Committee
SW	southwest
SWR	service water reservoir
TEDE	total effective dose equivalent
TLD	thermoluminescent dosimeter

TSC	technical support center
UFSAR	updated final safety analysis report
UHF	ultra-high frequency
UHS	ultimate heat sink
ULF	ultra-low frequency
USACE	U.S. Army Corps of Engineers
USBR	United States Bureau of Reclamation
USGS	United States Geological Survey
VCU	Virginia Commonwealth University
VDEM	Virginia Department of Emergency Management
VDGIF	Virginia Department of Game and Inland Fisheries
VDH	Virginia Department of Health
VSP	Virginia State Police
VT	Virginia Polytechnic Institute and State University
WHTF	waste heat treatment facility
ZPA	zero period acceleration

1. INTRODUCTION

Dominion Nuclear North Anna, LLC (Dominion or the applicant), filed an application with the U.S. Nuclear Regulatory Commission (NRC), docketed on October 23, 2003, for an early site permit (ESP) for a site the applicant designated as the North Anna ESP site. The proposed site is located near Lake Anna in Louisa County, Virginia, approximately 40 miles north-northwest of Richmond, Virginia.

The Dominion ESP application includes the site safety analysis report (SSAR), which describes the safety assessment of the site, as required by 10 CFR 52.17, "Contents of Applications." The public may inspect copies of this document via the Agencywide Documents Access and Management System (ADAMS) using ADAMS Accession No. ML032731517.[3] Dominion subsequently revised the application to address requests from the NRC staff for additional information. The applicant submitted SSAR Revision 5 (ADAMS Accession No. ML052150226) to the Commission by letter dated July 25, 2005. Throughout the course of the review, the staff requested that the applicant submit additional information to clarify the description of the North Anna site. Based on SSAR Revision 5, the staff issued NUREG-1835, "Safety Evaluation Report for an Early Site Permit (ESP) at the North Anna ESP Site," in September 2005.

In NUREG-1835, the staff documented its review of the site seismology, geology, meteorology, and hydrology, as well as the hazards to a nuclear power plant that could result from manmade facilities and activities on or in the vicinity of the site. NUREG-1835 also documented the staff's assessment of the risks of potential accidents that could occur as a result of the operation of a nuclear plant or plants at the site and evaluated whether the site could support adequate physical security measures for a nuclear power plant or plants. In NUREG-1835, the staff also evaluated whether the applicant's quality assurance measures are equivalent in substance to the measures discussed in Appendix B, "Quality Assurance Criteria for Nuclear Power Plants and Fuel Reprocessing Plants," to Title 10, Part 50, "Domestic Licensing of Production and Utilization Facilities," of the *Code of Federal Regulations* (10 CFR Part 50). The staff also evaluated the adequacy of the applicant's program for compliance with 10 CFR Part 21, "Reporting of Defects and Noncompliance." Finally, the staff reviewed the proposed major features of the emergency plan that Dominion would implement if a new reactor(s) is eventually constructed at the ESP site. The NRC would need to review the complete and integrated emergency plan in a separate licensing proceeding.

---

[3]ADAMS is the NRC's information system that provides access to all image and text documents that the NRC has made public since November 1, 1999, as well as bibliographic records (some with abstracts and full text) that the NRC made public before November 1999. Documents available to the public may be accessed via the Internet at http://www.nrc.gov/reading-rm/adams/web-based.html. Documents may also be viewed by visiting the NRC's Public Document Room at One White Flint North, 11555 Rockville Pike, Rockville, Maryland. Telephone assistance for using Web-based ADAMS is available at 800-397-4209 between 8:30 a.m. and 4:15 p.m., eastern standard time, Monday through Friday, except Federal holidays. The staff is also making this safety evaluation report available on the NRC's new reactor licensing public Web site at http://www.nrc.gov/reactors/new-licensing/esp/north-anna.html.

In a letter dated October 24, 2005 (ADAMS Accession No. ML052980117), Dominion notified the NRC that it had conducted additional evaluations of cooling water alternatives for a potential third nuclear reactor at its North Anna site. Based on those evaluations, Dominion decided to modify its approach for cooling a third unit from the base case currently described in the North Anna ESP application. The revised approach would reduce both thermal impacts and water consumption associated with Lake Anna. The applicant is taking this action partly in response to concerns that State regulatory bodies and local residents have raised. Under the revised approach, the proposed Unit 3 would employ a closed-cycle cooling system that would not use the 3400-acre waste heat treatment facility to dissipate waste heat. In addition, the evaporative loss of water from Lake Anna associated with cooling a third unit would be reduced from that previously considered.

In a letter dated November 22, 2005 (ADAMS Accession No. ML053260619), Dominion notified the NRC that it had selected the General Electric economic and simple boiling-water reactor (ESBWR) design for the preparation of a combined license (COL) application, thereby increasing the power level of the proposed Units 3 and 4 (hereafter referred to as Units 3 and 4) from 4300 megawatts thermal (MWt) to 4500 MWt. Dominion stated that the revised cooling approach will allow operation of an ESBWR at its full proposed design power level of 4500 MWt without thermal impact on the lake.

On January 13, 2006 (ADAMS Accession No. ML060250396), as described in Dominion letters dated October 24 and November 22, 2005, Dominion submitted a supplement discussing the modified approach for cooling a potential third unit at the North Anna ESP site. In addition, it also adjusted the North Anna ESP application plant parameter envelope (PPE) to reflect an increased core thermal power value of 4500 MWt (and corresponding estimated electrical output). Dominion stated that a future Revision 6 of the North Anna ESP application would ultimately incorporate the information contained in the supplement.

On April 13, 2006, Dominion submitted Revision 6 (ADAMS Accession No. ML061180180) to its application, which included a revised SSAR and environmental report. In Revision 6 to the North Anna ESP application, Dominion proposed (1) changing its approach for cooling Unit 3 from the once-through cooling system, as described in previous versions of the SSAR, to a closed-cycle system and (2) increasing the maximum power output per unit from 4300 MWt to 4500 MWt for Units 3 and 4. Under the revised cooling system approach, Unit 3 would use a closed-cycle, combination wet and dry cooling system. The proposed increase in power level corresponds to the revision of the designed maximum power (4,500 MWt) of an ESBWR, one of the reactor designs included in the PPE and evaluated by the NRC in the FSER.

The staff, in its review of Revision 6 of the application, requested that Dominion provide additional information. By letters dated June 21 and July 31, 2006, Dominion submitted Revisions 7 and 8 of the application (ADAMS Accession Nos. ML061870030 and ML062140009), addressing the staff's requests for additional information (RAIs). By letter dated September 12, 2006, Dominion submitted Revision 9 of the application (ADAMS Accession No. ML062580096). In this revision, Dominion decided to reduce the bounding value for tritium activity release (associated with the ACR-700 design) to ensure that the tritium concentration in liquid effluents resulting from normal operation is less than both the 10 CFR Part 20 limit and the EPA drinking water standards.

The staff decided to issue Supplement 1 to the FSER to evaluate the proposed cooling design change to Unit 3, the increase in the power level from 4300 MWt to 4500 MWt for Units 3 and 4, and changes to the application in Revisions 6, 7, 8, and 9.

This supplement includes the NRC staff's analysis of the safety aspects of constructing and operating a closed-cycle cooling system for Unit 3 and the increase in power for Units 3 and 4. The staff has identified, in Appendix A to this supplement, certain site-related items that an applicant will need to address at the COL or construction permit stage, should it desire to construct one or more new nuclear reactors on the North Anna ESP site. The staff determined that these items do not affect the staff's regulatory findings at the ESP stage and are more appropriately addressed at later stages in the licensing process. In addition, Appendix A to this supplement identifies the proposed permit conditions that the staff recommends the Commission impose, should the NRC issue an ESP to the applicant.

This supplement contains the sections and/or chapters of the FSER that are affected by the changes in the cooling design and the increase in the power level. The supplement section numbers may not be consecutive, but match the FSER sections that are being revised. Sections 2.3 and 2.4 of this supplement contain the evaluation of the impact of changes in the cooling design and increase in power. Chapters 11 and 15 of this supplement replace Chapters 11 and 15 of the FSER (NUREG-1835). The staff has revised Appendix A and provides it in its entirety for clarity and ease of inclusion in any ESP that the NRC may issue.

Appendix A to this FSER supplement lists the site characteristics, permit conditions, COL action items, and the bounding parameters that the staff is recommending that the Commission include in any ESP that the NRC might issue for the proposed site. Appendix B to this FSER details a chronology of the principal actions and correspondence related to the staff's review of the ESP application for the North Anna site. Appendix C lists the references for this FSER, Appendix D lists the principal contributors to this report, and Appendix E will include a copy of the Advisory Committee on Reactor Safeguards report.

The NRC has made the application and other pertinent information and materials available for public inspection at the NRC's Public Document Room at One White Flint North, 11555 Rockville Pike, Rockville, Maryland. The application and this SER are also available at the Louisa County Public Library, 881 Davis Highway, Mineral, Virginia, as well as on the NRC's new reactor licensing public Web site at http://www.nrc.gov/reactors/new-licensing/esp/north-anna.html.

2. SITE CHARACTERISTICS

2.3 Meteorology

Section 2.3 of NUREG-1835 describes in detail the meteorological setting of the North Anna ESP site, technical information contained in the site safety analysis report (SSAR), requests for additional information (RAIs) and their resolution, the regulatory basis for the staff safety evaluation, the staff's technical analyses, including independent verification by the staff of the meteorological site characteristics in Appendix A of this SER supplemental, and the staff's safety conclusions. This section of the SER supplement contains the staff's evaluation of the safety aspects of meteorological conditions at the North Anna ESP site affected by changes to the proposed Unit 3 normal plant cooling system design through Revision 9 of the SSAR.

In Revision 9 of the SSAR, the applicant stated that the proposed Unit 3 normal heat dissipation system would use a closed-cycle, dry and wet hybrid cooling system for cooling the circulating water system. A separate service water cooling system would use a closed-cycle wet cooling tower to dissipate waste heat from auxiliary heat exchangers not cooled by the plant circulating water system. This is a change from earlier versions of the SSAR, which stated that the proposed Unit 3 would use a once-through cooling system that would withdraw water from the North Anna Reservoir, circulate it through the condensers, and return the water to the reservoir via the waste heat treatment facility (WHTF).

Revision 9 of the SSAR offers no change in the proposed Unit 4 normal plant cooling system. The proposed Unit 4 would use a closed-cycle cooling system with dry cooling towers.

The only meteorological section affected in this SER supplement is Section 2.3.2, Local Meteorology.

2.3.2 Local Meteorology

2.3.2.1 Technical Information in the Application

In Section 2.3.2 of the SSAR, the applicant addressed, among other items, the potential influence of construction and operation of a nuclear power plant or plants falling within the applicant's plant parameter envelope (PPE) on local meteorological conditions that might in turn adversely impact the plant or plants or the associated facilities. The applicant also provided a topographical description of the site and its environs.

In Revision 9 of SSAR Section 2.3.2.3, the applicant stated that the increase in maximum daily surface water temperature in Lake Anna resulting from operation of the proposed Unit 3 cooling tower system would be negligible and would not impact the ongoing moderation of temperature extremes and alterations of wind patterns by the lake. The applicant also stated that the increase in the formation of fog induced by the discharge of cooling water to the lake because of the operation of the proposed Units 3 and 4 would be negligible.

In Revision 9 of SSAR Section 2.3.2.3, the applicant also stated that the operation of the wet cooling towers for the proposed Unit 3 may result in moisture deposition in the immediate vicinity of the towers because of drift and condensation of vapor near the discharge at the top of the towers. In addition, periodic fogging may occur around and downwind of the towers when atmospheric conditions are conducive to fog formation. The convective and conductive heat losses to the atmosphere resulting from the operation of the proposed Unit 4 closed-loop dry tower system could also result in localized increases in overall ambient temperature. The applicant concluded that it would consider the potential impact on the design or operation of the proposed unit(s) from any increase in the local ambient air temperature or moisture content induced by the cooling tower as part of detailed engineering.

In its letter dated March 2, 2006, the staff asked the applicant whether large-scale cut and fill activities would be needed to accommodate the additional land area used for the wet and dry cooling tower system for Unit 3. In its response letter dated April 13, 2006, the applicant revised Section 2.3.2.4 of the SSAR to define better the topography in the area of the defined ESP plant footprint and the necessary cut and fill activities in the proposed cooling tower area. The applicant stated that, should additional units be constructed, a portion of the currently undeveloped area of the ESP site would be cleared of existing vegetation and subsequently graded to accommodate the proposed units and the ancillary structures. No large-scale cut and fill activities would be needed in the area of the defined ESP site footprint to accommodate the proposed units since much of the area to be developed is already relatively level. Undulating surfaces in the area of the planned cooling towers would be leveled to accommodate the towers. Therefore, the applicant expects that terrain modifications associated with development of the ESP facility would be limited to the existing North Anna Power Station (NAPS) site and would not impact terrain features around the lake and valley or significantly alter the site's existing gently undulating surface, which is characteristic of the Piedmont region of Virginia.

2.3.2.2 Regulatory Evaluation

NUREG-1835 sets forth detailed regulatory evaluation of the application. This section of the ESP supplement focuses only on the change in approach to the proposed Unit 3 normal cooling presented in Revision 9 of the application.

Section 1.8 of the SSAR presents a detailed discussion of the applicant's conformance to NRC regulations and regulatory guidance. The applicant identified the applicable regulations as General Design Criterion (GDC) 2, "Design Bases for Protection Against Natural Phenomena," in Appendix A to 10 CFR Part 50, 10 CFR 100.20(c), and 10 CFR 100.21(d). The applicant identified the applicable regulatory guidance as Regulatory Guide (RG) 1.70, "Standard Format and Content of Safety Analysis Reports for Nuclear Power Plants - LWR Edition," and Review Standard (RS)-002, "Processing Applications for Early Site Permits." The staff reviewed this portion of the application in accordance with the guidance identified by the applicant to determine if the application complies with the identified regulations, with the exception that an ESP applicant need not demonstrate compliance with the GDC with respect to site meteorology.

2.3.2.3 Technical Evaluation

In versions of the SSAR preceding Revision 9, the applicant noted that it was not possible to predict with certainty the warm air transport and dispersion from the proposed Unit 4 dry cooling tower to specific plant features because the design of the plant was not known at the ESP stage. Therefore, the staff determined that the potential impact of the proposed Unit 4 dry cooling towers on the design and operation of the ESP facility should be considered as part of detailed engineering and receive further evaluation at the time of the COL application. This became **COL Action Item 2.3-1**.

In Revision 9 of SSAR Section 2.3.2.3, the applicant stated that it would consider as part of detailed engineering the potential impact on the design or operation of the proposed unit(s) from any increase in the local ambient air temperature or moisture content induced by the cooling towers for Unit 3 or 4. Since the specific layout and design of the ESP facility is not now known, the staff agrees that it is not possible to predict accurately the impact of either the Unit 3 or Unit 4 cooling tower plumes on specific plant features. The staff has determined that the COL or construction permit (CP) applicant also needs to consider the potential impact of moisture and salt deposition resulting from drift and condensation on plant design and operation as part of detailed engineering. Based on this submission, the staff is revising COL Action Item 2.3-1, which now provides that the COL or CP applicant should consider as part of detailed engineering the potential impact on the design or operation of the proposed unit(s) of any cooling-tower-induced local increase in (1) ambient air temperature, (2) ambient air moisture content, or (3) moisture and salt deposition.

In connection with revised COL Action 2.3-1, the staff finds that any effect cooling towers might have on local meteorological conditions can be treated as a design issue because such impacts are unlikely to have more than a minor effect on plant design as demonstrated by currently operating plants with cooling towers. Accordingly, such effects will be appropriately characterized and will not represent a potential threat posing an undue risk to Units 3 and 4.

Because of the limited and localized nature of the expected terrain modifications associated with the development of the ESP facility, including the proposed Unit 3 and 4 cooling tower systems, the staff finds that these terrain modifications, along with the resulting plant structures and associated improved surfaces (except for the proposed Unit 3 and 4 cooling towers as described above), will not have enough impact on local meteorological conditions to affect plant design and operation.

2.3.2.4 Conclusions

As described above, the applicant has presented and substantiated information on local meteorological and topographic characteristics of importance to the safe design and operation of a nuclear power plant or plants falling within the applicant's PPE that might be constructed on the proposed site. The staff has reviewed the information provided and, for the reasons given, concludes that the applicant's identification and consideration of the meteorological and topographical characteristics of the site and the surrounding area meet the requirements of 10 CFR Part 100, 10 CFR 100.20(c), and 10 CFR 100.21(d) and are sufficient to determine the acceptability of the site, with **COL Action Item 2.3-1**. The staff's conclusion relative to meteorology as put forth in NUREG-1835 remains valid for Revision 9 of the SSAR.

2.4 Hydrology

NUREG-1835 describes in detail the hydrologic setting of the North Anna ESP site, technical information contained in the SSAR, RAIs and their resolution, the regulatory basis for the staff safety evaluation, the staff's technical analysis, including independent verification of the water budget by the staff, and the staff's safety conclusions. This SER supplement contains the staff's evaluation of safety-related aspects at the North Anna ESP site affected by changes proposed by the applicant to the design of the Unit 3 normal plant cooling system through Revision 9 of the SSAR.

In Revision 9 of the SSAR, the applicant stated that the proposed Unit 3 normal heat sink would use a closed-cycle, dry and wet hybrid cooling tower system for cooling the circulating water. A separate service water cooling system would use a closed-cycle wet cooling tower to dissipate waste heat from auxiliary heat exchangers not cooled by the plant circulating water system. The applicant stated that the maximum instantaneous makeup water flow rate for Unit 3 wet towers would be 49.6 cfs, which would be withdrawn from Lake Anna. During normal operation, a maximum instantaneous blowdown discharge of 12.4 cfs from the wet towers would be discharged to Lake Anna via the WHTF.

Revision 9 of the SSAR offers no change in the proposed emergency cooling systems for Units 3 and 4. Both of these emergency cooling systems would use mechanical draft cooling towers over a buried water storage basin, if the selected reactor design needs an independent water source for its ultimate heat sink (UHS).

2.4.1.1 Technical Information in the Application

In Revision 9 of the SSAR, Section 2.4.1, the applicant changed the proposed design of the circulating water cooling system for Unit 3 from a once-through cooling system to a closed-cycle, dry and wet hybrid cooling tower system. The applicant proposed that makeup water would be supplied from Lake Anna and blowdown would be discharged to Lake Anna via the WHTF. The applicant stated that the plant service water cooling system for the proposed Unit 3 would use a separate wet cooling tower system.

The Unit 4 normal plant cooling system, which would use closed-cycle, dry cooling towers, remains unchanged. The applicant estimated the makeup water requirement for the Unit 4 dry towers to be about 1 gpm (0.002 cfs), which would also be supplied from Lake Anna using separate pumps located inside the new intake structure. The service water cooling system for Unit 4 would use separate dry cooling towers.

In Revision 9 of the SSAR, Section 2.4.7, the applicant stated that the proposed Units 3 and 4 separate the normal and emergency cooling water systems. The applicant attested that there are no interconnections or interreliance between normal and emergency cooling systems. The applicant also stated that the proposed normal cooling systems for Units 3 and 4 are reliable and would not be affected by ice formation in Lake Anna.

In Revision 9 of the SSAR, Section 2.4.7, the applicant described the conditions needed for formation of frazil ice, including a water temperature below 32 °F, a rate of supercooling greater than 0.018 °F per hour, turbulent flow with flow velocity of approximately 2 fps, and an absence of surface ice sheet. The water temperature at the NAPS intakes during the winter months has historically been above freezing as indicated by data obtained by Virginia Power as part of its thermal monitoring program. Even in the presence of turbulence resulting from winds, with the existing units operating and discharging warm cooling water to the lake, frazil ice would not be expected to form at the intakes for the proposed units. However, in case Units 1 and 2 do not operate for a prolonged period during a severe winter, frazil ice could form near the new intakes as a result of supercooling. The applicant stated that the new intakes would be designed with an intake velocity of 1 fps or less to reduce turbulence-induced conditions for frazil ice formation. The applicant also stated that under extreme conditions, with Units 1 and 2 not in operation, formation of frazil ice near the new intakes would not affect safety-related facilities adversely since the UHS would still be available with its independent water storage facility to shut down and maintain the plant in a safe mode.

In Revision 9 of the SSAR, Section 2.4.10, the applicant stated that it would provide rip-rap protection of slope embankment at the makeup water intake location for the new units to protect against wave erosion, although the intake is not a safety-related facility.

In Revision 9 of the SSAR, Section 2.4.11, the applicant stated that Lake Anna would provide the makeup water needed for cooling towers for the new units. The design of the intakes would be based on an elevation of 242 ft mean sea level (MSL), the plant shutdown low-water surface elevation in Lake Anna for all units including the proposed Units 3 and 4, with sufficient margin to ensure safe operation during low-water events.

The applicant updated its water budget analysis to reflect the proposed changes to the Unit 3 normal plant cooling approach. The applicant's minimum calculated water surface elevations in Lake Anna for the existing units (only existing Units 1 and 2 running at a plant capacity factor of 93 percent) and proposed (existing units as before, new Unit 3 operating with a 96 percent capacity factor, and Unit 4 operating with no need for makeup water and no blowdown discharge) for the assumed scenarios are 245.1 and 244.2 ft MSL, respectively.

2.4.1.2 Regulatory Evaluation

NUREG-1835 sets forth a detailed regulatory evaluation of the application. This section of the ESP supplement focuses only on the change in approach to Unit 3 normal cooling proposed in Revision 9 of the application.

Section 1.8 of the SSAR presents a detailed discussion of the applicant's conformance to NRC regulations and regulatory guidance. The applicant identified the applicable regulations as GDC 2 in Appendix A to 10 CFR Part 50, 10 CFR 52.17(a), 10 CFR 100.20(c), 10 CFR 100.23(c), Section IV(c) of Appendix S to 10 CFR Part 50, and 10 CFR 100.21(d). The applicant identified the applicable regulatory guidance as RG 1.70 and RS-002. The staff reviewed this portion of the application for conformance with the applicable regulations and considered the corresponding regulatory guidance identified above, with the exception that an ESP applicant need not demonstrate compliance with the GDC with respect to the hydrological description of the site.

2.4.1.3 Technical Evaluation

This section describes the staff's review of the applicant's submission in Revision 9 of the SSAR. This review focused only on safety-related aspects that may be affected by the change in approach to proposed Unit 3 normal cooling.

In Revision 9 of the SSAR, the applicant changed the approach for normal plant cooling design and proposed that the design of the Unit 3 circulating water cooling system would use a dry and wet hybrid cooling tower system. The staff reviewed the applicant's revised water budget for the proposed approach for normal cooling for Unit 3. Since the applicant did not propose any change to the emergency-cooling-related function of the UHS, the staff safety review focused on two key parameters that can indicate the potential for frequent and sudden reliance on the Unit 3 UHS. These are (1) the frequency of lake water surface elevation dropping to the shutdown level of 242 ft MSL when all units are in operation and (2) the velocity with which the lake water surface elevation drops to the shutdown level of 242 ft MSL. The staff estimated the frequency of reduced water surface elevations in Lake Anna by examining data collected for the 25-year period between 1978 and 2003.

Using PPE values for operation of ESP Unit 3 and the historical record for other conditions, the staff conservatively estimated that the minimum level of Lake Anna during the 2001–2002 critical period of draught (the two consecutive driest years from 1921-2004) would have been 242.9 ft MSL. This elevation value is above the shutdown elevation (242 ft MSL) for both ESP Unit 3 and NAPS Units 1 and 2. The staff also determined that the rate of drop in water surface elevation of Lake Anna in the presence of new Units 3 and 4 will be gradual and will allow sufficient time for safe shutdown of Unit 3. The staff used a natural evaporation rate of 5.6 in./mo from the literature, a Unit 3 evaporation rate of 1.9 in./mo based on the PPE value, an existing NAPS units' evaporation of 5.6 in./mo based on a conservative assumption that all heat load from the existing units would go towards latent heat of vaporization, and an equivalent reduction in the Lake Anna water level of 1.5 in./mo resulting from a minimum release of 20 cfs over the dam. Summing these amounts results in a maximum effective evaporation rate of 14.6 in./mo. Using this conservative evaporation rate, the staff estimated that it would take 49 days for water surface elevation in Lake Anna to drop from 244 ft MSL to 242 ft MSL. Therefore, the staff concludes that water surface elevation in Lake Anna does not fall rapidly and that sufficient time will be available to plant operators before the low water surface elevation shutdown threshold is reached to plan a shutdown of the proposed Unit 3 without endangering its safety, even under severe drought conditions. The staff concludes, therefore, that the safety conclusions in NUREG-1835 do not change as a result of the applicant's submission in Revision 9 of the SSAR.

The maximum makeup water for the wet cooling portion of the Unit 3 cooling system is 49.6 cfs, which would be withdrawn from Lake Anna. Based on this information, the staff is withdrawing COL Action Item 2.4-3, which was based on once-through cooling Unit 3 and is no longer relevant to the staff's conclusions. Instead the staff is proposing a new controlling PPE for makeup water for the Unit 3 cooling system, the value of which would be a maximum makeup water withdrawal rate of 49.6 cfs. The staff had also previously identified three other controlling PPE values with respect to the Unit 3 once-through cooling system, namely, maximum cooling water flow rate of 2540 cfs, maximum inlet temperature of 95 degrees F, and maximum temperature rise of 18 degrees F. With the closed cycle wet and dry cooling system proposed by the applicant, the parameters related to maximum inlet temperature and the temperature rise are no longer necessary as bounding parameters. For the same reason, the maximum makeup water bounding value replaces that of the maximum cooling water flow rate. Therefore, these parameters are removed from Appendix A, Section A.4.

Buried storage reservoirs with their associated mechanical draft cooling towers will provide the UHS if the selected reactor design for the proposed Units 3 and 4 needs an independent source of water. These reservoirs are completely independent of Lake Anna except for their initial filing and periodic replenishment.

2.4.1.4 Conclusions

As described above, the applicant has provided information pertaining to the hydrology of the site in Revision 9 of the SSAR. Therefore, the staff concludes that, with the noted bounding PPE values, the applicant has met the requirements regarding hydrology in 10 CFR 52.17(a), 10 CFR 100.20(c), 10 CFR 100.23(c), and 10 CFR 100.21(d) with respect to the matters discussed above. The staff concludes that, subject to the bounding PPE values discussed above, the staff's conclusions relative to hydrology as put forth in NUREG-1835 remain valid for Revision 9 of the SSAR.

11. RADIOLOGICAL EFFLUENT RELEASE DOSE CONSEQUENCES FROM NORMAL OPERATIONS

11.1 Source Terms

The U.S. Nuclear Regulatory Commission (NRC) staff has reviewed the information on radiological dose consequences caused by gaseous and liquid effluents that may be released from normal operation of the plant that was provided by reference in Site Safety Analysis Report (SSAR) Section 2.3.5.1, SSAR Tables 1.3-7 and 1.3-8, and included in the Environmental Report (ER), Section 3.1 (Table 3.1-9), and Section 5.4 (Tables 5.4-9, 5.4-10, and 5.4-11) of the Dominion Nuclear North Anna, LLC (Dominion or the applicant), early site permit (ESP) application, Revision 9. The purpose of the evaluation is to determine whether site characteristics are such that radiation doses to members of the public would be within applicable regulatory requirements, as summarized in Table 11.1-1.

11.1.1 Technical Information in the Application

The applicant provided information on the radiological impacts on members of the public from gaseous and liquid effluents that would be generated as a normal byproduct of nuclear power operations. The applicant described the exposure pathways by which radiation and radioactive effluents can be transmitted to members of the public in the vicinity of the site. The estimates on the maximum doses to the public are based on the available data on the reactor designs being considered using the plant parameter envelope (PPE) approach in which the bounding liquid and gaseous radiological effluents were used in assessing impacts on the public. The applicant evaluated the impact of these doses by comparing them to applicable regulatory limits.

Using the PPE approach, Dominion provided a list of fission and activation products that may be released in liquid and gaseous effluents from the postulated two new units. The applicant evaluated the impacts from releases and direct radiation by considering the probable pathways to individuals, populations, and biota near the proposed new units. The applicant also calculated the highest dose from the major exposure pathways for a given receptor.

If built, the postulated two new units at the North Anna ESP site would release liquid effluents into the waste heat treatment facility (WHTF) and Lake Anna through the discharge canal used for the currently operating units. The applicant considered the following liquid pathways: ingestion of aquatic food; ingestion of drinking water; exposure to shoreline sediment; and exposure to water through boating and swimming. Exposures associated with crop and pasture irrigation were not considered because the use of water from Lake Anna was deemed negligible for this purpose.

Dominion also considered gaseous pathways, including external exposure to the airborne plume, external exposure to contaminated ground, inhalation of airborne activity, and ingestion of contaminated agricultural products, in its application.

Table 11.1-1

Staff's Summary of 10 CFR Part 50 Appendix I Dose Objectives and 40 CFR Part 190 Environmental Dose Standards

Regulation	Type of Effluent	Pathway	Organ	Dose Limit (mrem/yr per unit)
10 CFR Part 50, Appendix I *	Liquid	all	total body	3
		all	any organ	10
	Gaseous	all	total body	5
		all	skin	15
	Radioiodines & Particulates	all	any organ	15
	Gaseous	gamma air dose	n/a	10***
		beta air dose	n/a	20***
40 CFR Part 190 **	all	all	total body	25#
	all	all	thyroid	75#
	all	all	any other organs	25#

Notes:
*   Appendix I dose objectives are defined for the maximally exposed individual (MEI).
**   Dose limits are defined for any real member of the public. Under NRC requirements, this standard is implemented under 10 CFR Part 20.1301(e).
***   Air doses are expressed in mrad/year instead.
#   40 CFR Part 190 dose limits are for the entire site and apply to all operating units.

The applicant calculated the dose to the maximally exposed individual (MEI) from both the liquid and gaseous effluent release pathways, and calculated a collective whole body dose for the population within 50 miles (80 km) of the North Anna ESP site.

11.1.2 Regulatory Evaluation

NRC regulations require that applicants for an ESP address the characteristics of the proposed site that could affect radiation doses to a member of the public from radiological effluents. In SSAR Section 1.8.1, the applicant identified the applicable NRC regulations as Title 10, Section 52.17(a)(1)(iv), of the *Code of Federal Regulations* (10 CFR 52.17(a)(1)(iv)). Specifically, this regulation states that an ESP application should describe the anticipated maximum levels of radiological effluents that each facility will produce during normal operations. Furthermore, 10 CFR 100.21(c)(1) requires that site atmospheric dispersion parameters be established such that radiological effluent release limits associated with normal operation from the type of facility proposed to be located at the site be met for any individual located off site. The staff reviewed this portion of the application for conformance with applicable regulations.

11.1.3 Technical Evaluation

During normal operation, small quantities of radiological materials are expected to be released to the environment through gaseous and liquid effluents from the plant.

11.1.3.1 Gaseous Effluents

The applicant calculated the estimated dose to a hypothetical maximally exposed member of the public from gaseous effluents using radiological exposure models based on Regulatory Guide (RG) 1.109, Revision 1, "Calculation of Annual Doses to Man from Routine Releases of Reactor Effluents for the Purpose of Evaluating Compliance with 10 CFR Part 50, Appendix I," issued October 1977; the GASPAR II computer program (NUREG/CR-4653; "GASPAR II - Technical Reference and User Guide," March 1987); and RG 1.111, Revision 1, "Methods for Estimating Atmospheric Transport and Dispersion of Gaseous Effluents in Routine Releases from Light-Water-Cooled Reactors," issued July 1977. Section 2.3.5 of the SSAR discusses the derivation of the atmospheric dispersion parameters, and presents the specific values of the dispersion parameters used in the applicant's radiological dose assessment.

Dominion calculated the gaseous pathway doses to the MEI using the GASPAR II program at the nearest site boundary, nearest vegetable garden, nearest residence, and nearest meat cow. The milk exposure pathway was not considered because there are no reported cows or goats used for milk production in the near vicinity of the site (within a 5-mile radius). In Table 1.3-8 of the SSAR and Table 5.4-7 of the ER, the applicant provided an estimate of the radiological source term associated with gaseous effluents that may be released from normal operation of the plant. The applicant developed estimates of gaseous radioactive effluent concentration levels based on a composite of the highest activity levels of individual radionuclides it anticipates to be released from alternative reactor designs under consideration. These releases reflect composite estimates based on the advanced boiling-water reactor (ABWR), Advanced Plant 1000 (AP1000), Atomic Energy of Canada Advanced CANDU Reactor (ACR-700), and the economic and simplified boiling-water reactor (ESBWR) reactor designs, with the ABWR gaseous effluent source term scaled up to a power rating of 4300 MWt from the certificated design of 3926 MWt, and the ESBWR source term increased by a 25 percent margin. The ESBWR is rated at 4500 MWt. This approach results in a slight increase in the assumed release rate of those radionuclides for which the ABWR and ESBWR designs were assumed to be bounding.

The gaseous effluent releases are used to estimate doses to the MEI. Tables 5.4-3 through 5.4-5 of the environmental report include other inputs to the GASPAR II program, including meat and vegetable production rates, atmospheric dispersion and ground deposition factors, receptor locations, and the assumed consumption rates of food products by the MEI.

Tables 5.4-9 and 5.4-10 of the environmental report present the gaseous pathway doses to the MEI calculated by the applicant. The applicant calculated gaseous pathway doses to the MEI, including a maximum annual dose to the total body doses of 0.014 milliSievert (mSv) (1.4 millirem (mrem)) at the nearest residence from the plume; 0.021 mSv (2.1 mrem) at the nearest site boundary from the plume; 0.0031 mSv (0.31 mrem) at the nearest site boundary from inhalation (teen); 0.011 mSv (1.1 mrem) at the nearest garden (child); 0.002 mSv (0.2 mrem) at the nearest residence from inhalation (adult and teen); and 0.00079 mSv (0.079 mrem) for the nearest cow meat location. Dominion calculated a maximum annual dose to the thyroid of 0.13 mSv (13 mrem) for the child based on the location of the nearest garden. These calculated doses satisfy the 10 CFR Part 50 Appendix I Sections II.B and II.C dose objectives for the MEI. The staff performed an independent evaluation of gaseous effluent pathway doses using the GASPAR II code and the applicant's input data and calculated similar results. See applicable dose criteria in Table 11.1-1.

The applicant also estimated bounding gaseous effluent radionuclide concentrations for receptors assumed to be located at the exclusion area boundary (EAB) in demonstrating compliance with gaseous effluent release concentration limits of 10 CFR Part 20, "Standards for Protection Against Radiation." The effluent concentration limits are contained in Table 2 (Column 1) of Appendix B to Part 20, under "Annual Limits on Intake (ALIs) and Derived Air Concentrations (DACs) of Radionuclides for Occupational Exposure; Effluent Concentrations; Concentrations for Release to Sewerage." The applicant's results are presented in Table 5.4-7 of the environmental report and fall below the concentration limits of 10 CFR Part 20, Appendix B. The staff performed an independent evaluation of the estimated effluent concentration levels using the applicant's data for the source term and χ/Q at the EAB and found similar results.

Under the requirements of 10 CFR 20.1301(e), the applicant also demonstrated compliance with the environmental radiation standards of the U.S. Environmental Protection Agency, under 40 CFR Part 190, "Environmental Radiation Protection Standards for Nuclear Power Operations." The applicant's results, based on the above doses, are therefore acceptably derived, and are presented in Table 5.4-11 of the environmental report for the MEI.

The applicant's results of 6.4 mrem/yr (0.064 mSv) for the whole body, 27 mrem/yr (0.27 mSv) for the thyroid, and 11 mrem/yr (0.11 mSv) to bone are smaller than the maximum doses specified in 40 CFR Part 190.10(a) of 25 mrem/yr whole body, 75 mrem/yr thyroid, and 25 mrem/yr any other organ. See Table 11.1-1.

Based on the above, the staff concludes that the applicant has provided a bounding assessment in demonstrating its capability to comply with the regulatory requirements in 10 CFR Part 20 "Standards for Protection Against Radiation," and Appendix I, "Numerical Guides for Design Objectives and Limiting Conditions for Operation to Meet the Criterion "As Low as is Reasonably Achievable" for Radioactive Material in Light-Water-Cooled Nuclear Power Reactor Effluents," to

10 CFR Part 50, "Domestic Licensing of Production and Utilization Facilities," given atmospheric dispersion parameters set forth in Section 2.3.5 of NUREG-1835.

11.1.3.2 Liquid Effluents

The applicant calculated the estimated dose to a hypothetical maximally exposed member of the public from liquid effluents using radiological exposure models based on RG 1.109 and the LADTAP II computer program (NUREG/CR-4013, "LADTAP II - Technical Reference and User Guide," April 1986).

Dominion calculated liquid pathway doses using the LADTAP II program for various activities, including eating fish and invertebrates assumed to be caught at the end of the discharge canal; drinking water from Lake Anna; boating and swimming; and using the shoreline for recreational purposes. In Table 1.3-7 of the SSAR and Table 5.4-6 of the ER, the applicant estimated the radiological source term associated with liquid effluents that may be released from normal operation of the plant. The applicant developed estimates of liquid radioactive effluent concentration levels based on a composite of the highest activity levels of individual radionuclides it anticipates to be released from alternative reactor designs under consideration. These releases reflect composite estimates based on the ABWR, AP1000, ACR-700, and the ESBWR reactor designs, with the ABWR liquid effluent source term scaled up to a power rating of 4300 MWt from the certificated design of 3926 MWt, and the ESBWR source term increased by a 25 percent margin. The ESBWR is rated at 4500 MWt. This approach results in a slight increase in the assumed release rate of those radionuclides for which the ABWR and ESBWR designs were assumed to be bounding.

The liquid effluent releases are used to estimate doses to the MEI. Tables 5.4-1 and 5.4-2 of the environmental report include other parameters used as input to the LADTAP II program, including effluent discharge flow rate, site-specific dilution flow rate, and transit time to receptor. The analysis assumed direct releases into the water body.

Tables 5.4-8 and 5.4-10 of the environmental report present the liquid pathway doses to the MEI calculated by the applicant. The applicant calculated liquid pathway doses to the MEI, including a maximum annual dose to the total body of 0.0081 mSv (0.81 mrem) for the adult. Dominion calculated a maximum annual dose to the thyroid of 0.0068 mSv (0.68 mrem) for the infant, and a maximum annual dose to the bone as 0.025 mSv (2.5 mrem) for the child. These calculated doses satisfy the 10 CFR Part 50 Appendix I, Section II.A dose objectives for the MEI. The staff performed an independent evaluation of liquid pathway doses using the LADTAP II code and the applicant's input data and found similar results. See applicable dose criteria in Table 11.1-1.

The applicant also estimated bounding liquid effluent radionuclide concentrations for receptors assumed to be located at the end of the discharge canal in demonstrating compliance with liquid effluent release concentration limits of 10 CFR Part 20. The effluent concentration limits are contained in Table 2 (Column 2) of Appendix B to Part 20. The applicant's results are presented in Table 5.4-6 of the environmental report and fall below the concentration limits of 10 CFR Part 20, Appendix B. The staff performed an independent evaluation of the estimated effluent concentration levels using the applicant's data for the source term and dilution factor and found similar results.

Under the requirements of 10 CFR 20.1301(e), the applicant also demonstrated compliance with the environmental radiation standards of the U.S. Environmental Protection Agency, under

40 CFR Part 190. The applicant's results, based on the above doses, are therefore acceptably derived, and are presented in Table 5.4-11 of the environmental report for the MEI.

The applicant's results of 6.4 mrem/yr (0.064 mSv) for the whole body, 27 mrem/yr (0.27 mSv) for the thyroid, and 11 mrem/yr (0.11 mSv) to bone are smaller than the maximum doses specified in 40 CFR Part 190.10(a) of 25 mrem/yr whole body, 75 mrem/yr thyroid, and 25 mrem/yr any other organ. See Table 11.1-1.

The staff concludes that the applicant has provided a bounding assessment in demonstrating its capability to comply with the regulatory requirements in 10 CFR Part 20 and Appendix I to 10 CFR Part 50.

11.1.4 Conclusions

As set forth above, the applicant provided information adequate to provide reasonable assurance that it will control, monitor, and maintain radioactive gaseous and liquid effluents from the ESP site within the regulatory limits described in 10 CFR Part 20, Appendix B, Table 2, as well as maintain them at levels that are in accordance with the effluent design objectives contained in Appendix I to 10 CFR Part 50, Sections II.A, II.B, and II.C. Under the requirements of 10 CFR 20.1301(e), the applicant also demonstrated compliance with the environmental radiation standards of the U.S. Environmental Protection Agency, under 40 CFR Part 190.

As further set forth above, the staff has independently confirmed the adequacy of the applicant's dose consequence calculations from normal operations. A combined license (COL) or construction permit (CP) applicant that references an ESP for the North Anna site should verify that the calculated radiological doses to members of the public from radioactive gaseous and liquid effluents for two new units which may be built at the North Anna site are bounded by the radiological doses included in the ESP application and reviewed by the NRC staff, as described above. This includes any changes made to address differences in reactor design used to calculate radiological doses (e.g., basis of the liquid and gaseous radiological source terms, and liquid effluent discharge flow rates and site-specific dilution flow rates). Exposures associated with crop and pasture irrigation were not considered because the use of water from Lake Anna was deemed to be negligible for this purpose. However, should local land-use information reveal that the use of water from Lake Anna becomes significant in irrigating crops and pastures, the COL or CP applicant should consider this pathway in the application and confirm that the associated doses are in compliance with applicable NRC criteria. In addition, detailed information on the solid waste management system used to process radioactive gaseous and liquid effluents will be necessary to reflect plant and site-specific COL design considerations. These items are addressed collectively in **COL Action Item 11.1-1.**

Based upon these considerations, the staff concludes that radiological doses to members of the public from radioactive gaseous and liquid effluents resulting from the normal operation of one or two new nuclear power plants that might be constructed on the proposed ESP site do not present an undue risk to the health and safety of the public. Therefore, the staff concludes, with respect to radiological effluent releases and dose consequences from normal operations, that appropriate long-term atmospheric dispersion coefficients have been established at the proposed site is acceptable for constructing one or two units falling within the applicant's bounding site-specific PPE, and that the site meets the relevant requirements of 10 CFRPart 52, "Early Site Permits; Standard Design Certifications; and Combined Licenses for Nuclear Power Plants," and 10 CFR Part 100, "Reactor Site Criteria."

15. ACCIDENT ANALYSES

15.1 Technical Information in the Application

In Chapter 15, "Accident Analyses," of the site safety analysis report (SSAR), the applicant analyzed the radiological consequences of design-basis accidents (DBAs) to demonstrate that new nuclear units could be located at the proposed early site permit (ESP) site without undue risk to the health and safety of the public, in compliance with the requirements of Title 10, Section 52.17, "Contents of Applications," of the *Code of Federal Regulations* (10 CFR 52.17) and 10 CFR Part 100, "Reactor Site Criteria." The applicant did not identify a particular reactor design to be considered for the proposed ESP site. Instead, the applicant developed a set of reactor DBA source term parameters using surrogate reactor characteristics. The applicant used these parameters in conjunction with site characteristics for accident analysis purposes to assess the suitability of the proposed ESP site. These plant parameters collectively constitute a plant parameter envelope (PPE).

The applicant developed a PPE using seven reactor designs (five water-cooled reactors and two gas-cooled reactors), though it used source terms for only three of these designs as inputs to its DBA analyses. The water-cooled reactors included in the PPE were (1) a version of the Westinghouse Advanced Plant 1000 (AP1000), (2) the certified General Electric (GE) Advanced Boiling-Water Reactor (ABWR), (3) the Atomic Energy of Canada Advanced CANDU Reactor (ACR-700), (4) a version of the GE Economic and Simple Boiling-Water Reactor (ESBWR), and (5) the Westinghouse-led International Reactor Innovative and Secure (IRIS) reactor. The ACR-700 is light-water cooled but heavy-water moderated. The two gas-cooled reactors are (1) the General Atomics Gas Turbine Modular Helium Reactor (GT-MHR) and (2) the Pebble Bed Modular Reactor (PBMR). The applicant stated that the PPE values are not intended to be limited to these reactor designs but rather to provide a broad overall outline of a design concept and to include other potential reactor designs if they fall within the parameter values provided in the PPE.

In selecting DBAs for dose consequence analyses, the applicant focused on three light-water reactors (LWRs), the certified ABWR, a version of the AP1000,[4] and a version of the ESBWR[5] to serve as surrogates. The applicant stated that it selected these three reactor designs because they are (or are based on) previously certified standard designs and have recognized bases for postulated accident analyses. Using source terms developed from these three designs, the applicant performed and provided radiological consequence analyses for the following DBAs:

- pressurized-water reactor (PWR) main steamline break
- PWR feedwater system pipe break

---

[4] As discussed later in this section, the applicant referenced a version of the AP1000 design available at the time it submitted its ESP application. Westinghouse subsequently revised the AP1000 design before the U.S. Nuclear Regulatory Commission (NRC) staff's issuance of a final safety evaluation report (SER) for the AP1000 design certification.

[5] The ESBWR considered by the applicant is based on Revision 1 of the ESBWR Design Control Document, Tier 2, submitted by GE in January 2006. The applicant increased the accident source terms by a factor of 1.25 to accommodate uncertainties because the NRC has not yet completed its design certification review.

- locked rotor accident
- reactor coolant pump shaft break
- PWR rod ejection accident
- BWR control rod drop accident
- failure of small lines carrying primary coolant outside containment
- PWR steam generator tube failure
- BWR main steamline break
- PWR and BWR loss-of-coolant accidents
- fuel-handling accident
- BWR cleanup water line break

The applicant presented the dose consequence assessment results in SSAR Chapter 15, "Accident Analyses." SSAR Table 15.4-1, "Summary of Design Basis Accident Doses," summarizes the postulated radiological consequences of the DBAs identified above at the proposed exclusion area boundary (EAB) and the low-population zone (LPZ) boundary. The potential doses set forth in the table would be within the radiological dose consequence evaluation factors set forth in 10 CFR 50.34(a)(1). The applicant provided the accident-specific source terms (release rates of radioactive materials from the ESP footprint (PPE values) to the environment) and resulting site-specific dose consequences for each DBA in Tables 15.4-3 through 15.4-31 of the SSAR.

In Request for Additional Information (RAI) 15.4-1, the staff noted that Westinghouse revised its atmospheric dispersion factors (χ/Q values) in the AP1000 design control document (DCD) since the applicant submitted the North Anna ESP application, and asked whether the applicant planned to use the updated values in revising its application. The applicant responded that it had elected not to update the ESP application to incorporate the latest χ/Q values in the AP1000 design certification. The applicant further stated that site-specific doses would be updated, as necessary, in any combined license (COL) or construction permit (CP) application, after selection of a specific reactor design.

In RAIs 15.4-2 and 15.4-3, the staff noted that SSAR Section 15.4 provides total effective dose equivalent (TEDE) values for the ABWR design, while the ABWR design is certified with the thyroid and whole body doses specified in 10 CFR Part 100. The staff asked the applicant to compare these doses. In its response, the applicant stated that it would revise the SSAR to include the thyroid and whole body doses from the ABWR DCD, in addition to the estimated TEDE values. The applicant incorporated this information into its application in Revision 3.

In RAI 15.4-4, the staff asked the applicant to provide references and explain the methodology it used to determine time-dependent activity releases for each DBA. The applicant provided the requested references. In its response, the applicant stated that the respective DCDs present the methodologies used for calculating time-dependent releases for the ABWR and AP1000. These methodologies are approved as acceptable in 10 CFR Part 52, Appendix A and Appendix D for the ABWR and AP1000, respectively.

In RAI 15.4-5, the staff asked the applicant to provide, for each DBA, the doses it used for the EAB and the LPZ for the AP1000 and the ABWR, as well as the ratios of site-specific χ/Q values to design certification χ/Qs used. In its response, the applicant stated that it would revise the dose tables in SSAR Section 15.4 to show the χ/Q values and doses from the AP1000 and ABWR DCDs, in addition to the ratios of site-specific χ/Q values to design certification χ/Q values. The applicant incorporated this information into its application in Revision 3.

In RAI 15.4-6, the staff asked the applicant to clarify whether the 0- to 2-hour EAB doses presented in the SSAR are for the 2-hour period with the greatest EAB doses and, if they are not, to provide the doses for the 2-hour period with the greatest EAB doses. In its response, the applicant stated that the greatest EAB dose occurs during the first 2 hours of the accident for all AP1000 accidents evaluated in SSAR Chapter 15, except for a loss-of-coolant accident (LOCA). As indicated in Section 15.6.5.3.8.1 of the AP1000 DCD, the period from 1 to 3 hours yields the greatest EAB dose for a LOCA. In view of the accident progression sequences for the designs used in the DBA dose assessment, the staff agrees with the applicant's conclusion. The applicant incorporated this information into its application in Revision 3.

In Supplemental RAI 1, dated May 12, 2006, the staff asked the applicant to provide the activity release for the period giving the highest 2-hour dose at the EAB for four ESBWR DBAs; (1) failure of small lines carrying primary coolant outside containment, (2) main steamline break (equilibrium activity), (3) LOCA, and (4) fuel-handling accident. In its response, the applicant stated that the maximum EAB dose occurs between 2 and 4 hours for the failure of small lines carrying primary coolant outside containment, the first 2 hours for the main steamline break and fuel-handling accidents, and between 2.6 and 4.6 hours for the LOCA. The applicant provided RAI Tables 1-1, 1-2, and 1-3 showing the activity releases for the periods giving the highest 2-hour dose at the EAB.

In Supplemental RAIs 2, 3, and 5, the staff asked the applicant to verify that the activity releases were correct, and to calculate the resulting doses for three ESBWR DBAs; (1) failure of small lines carrying primary coolant outside containment, (2) main steamline break (equilibrium activity), and (3) fuel-handling accident. In its response, the applicant changed the methodology for calculating all ESBWR DBA doses to the methodology used by the staff. Instead of applying χ/Q ratios to DCD doses, the ESBWR DBA doses are calculated directly based on the activity releases. The staff is currently reviewing ESBWR DBA doses provided by GE, the ESBWR vendor. The TEDE from an isotope for a given time period is calculated by adding the committed effective dose equivalent (CEDE) from inhalation and the effective dose equivalent (EDE) from external exposure. The CEDE is calculated by multiplying the isotopic activity released by the site χ/Q value, the breathing rate of the individual located offsite, and the effective inhalation dose conversion factor from Federal Guidance Report 11, "Limiting Values of Radionuclide Intake and Air Concentration and Dose Conversion Factors for Inhalation, Submersion, and Ingestion," issued 1988. The EDE is calculated by multiplying the isotopic activity by the site χ/Q value and the effective submersion dose conversion factor from Federal Guidance Report 12, "External Exposure to Radionuclides in Air, Water, and Soil," issued 1993. These methodologies follow the guidance in RG 1.183.

The applicant used the revised dose calculation methodology described above for all ESBWR DBAs, and produced the same doses as calculated by the staff. Revision 9 of the SSAR gives the revised doses in Tables 15.4-5d, 15.4-12b, 15.4-19b, 15.4-19c, 15.4-23b, 15.4-29, and 15.4-31.

In Supplemental RAI 4, the staff asked the applicant to provide a reference for the activity releases for the ESBWR LOCA presented in SSAR Table 15.4-23a. In its response, the applicant stated that it had obtained the releases from a formal correspondence from GE to Dominion dated March 1, 2006. The applicant provided a copy of the correspondence.

In Supplemental RAI 7, the staff asked the applicant to (1) explain why the SSAR analyzes the ABWR feedwater system pipe break (FLB) while the ABWR DCD does not analyze it and

(2) provide supporting information (activity release and dose calculation table) for the FLB with references for this information. The staff also asked the applicant to provide the supporting documentation for the ABWR cleanup waterline break (CLB) accident. In its response, the applicant stated that the ABWR DCD lists both feedwater and cleanup line break accidents even though it does not report the doses for the FLB, but simply refers the FLB to the CLB stating that the doses for the FLB bound those for the CLB. The ESBWR DCD evaluates both the FLB and the CLB, and the applicant added these two DBAs to Table 15.4-1 of the SSAR. For completeness, the applicant also added these two DBAs for the ABWR in the SSAR, Table 15.4-1. The applicant revised SSAR Table 15.4-1 in Revision 9, with a new note which states that the ABWR DCD indicates that the CLB bounds the doses for the FLB. The applicant also provided the supporting documentation for the ABWR CLB in SSAR, Revision 9, Tables 15.4-5a and 15.4-5b.

In Supplemental RAI 8, the staff asked the applicant to revise the SSAR to reflect any changes in activities and doses as a result of the above supplemental RAIs. The applicant has revised the SSAR accordingly in its Revision 9.

15.2 Regulatory Evaluation

In SSAR Section 1.8 and in SSAR Chapter 15, the applicant identified the following applicable NRC regulations and guidance regarding reactor accident radiological consequence analyses:

- 10 CFR 52.17

- 10 CFR Part 100

- 10 CFR 50.34, "Contents of Applications; Technical Information"

- Regulatory Guide (RG) 1.3, "Assumptions Used for Evaluating the Potential Radiological Consequences of a Loss of Coolant Accident for Boiling Water Reactors," issued June 1974

- RG 1.25, "Assumptions Used for Evaluating the Potential Radiological Consequences of a Fuel Handling Accident in the Fuel Handling and Storage Facility for Boiling and Pressurized Water Reactors," issued March 1972

- RG 1.145, "Atmospheric Dispersion Models for Potential Accident Consequence Assessments at Nuclear Power Plants," issued November 1982

- RG 1.183, "Alternative Radiological Source Terms for Evaluating Design Basis Accidents at Nuclear Power Reactors," issued July 2000

- NUREG-0800, "Standard Review Plan for the Review of Safety Analysis Reports for Nuclear Power Plants," issued July 1981

- TID-14844, "Calculation of Distance Factors for Power and Test Reactor Sites," issued March 1962

- Review Standard (RS)-002, "Processing Applications for Early Site Permits," issued May 3, 2004

The staff reviewed SSAR Section 1.8 and Chapter 15 for conformance with the applicable regulations and considered the corresponding guidance, as identified above. In its evaluation, the staff used the dose consequence evaluation factors found in 10 CFR 50.34(a)(1) that contribute to determining the acceptability of the site in accordance with 10 CFR 52.17(a)(1).

The regulations at 10 CFR 52.17(a)(1) require that ESP applications contain an analysis and evaluation of the major structures, systems, and components of the facility that bear significantly on the acceptability of the site under the radiological consequence evaluation factors identified in 10 CFR 50.34(a)(1). In addition, the ESP site characteristics must comply with the requirements of 10 CFR Part 100. The regulations at 10 CFR 50.34(a)(1)(ii)(D) require the following for a postulated fission product release based on a major accident:

- An individual located at any point on the boundary of the exclusion area for any 2-hour period following the onset of the postulated fission product release would not receive a radiation dose in excess of 25 rem TEDE.

- An individual who is located at any point on the boundary of the LPZ and who is exposed to the radioactive cloud resulting from the postulated fission product release (during the entire period of its passage) would not receive a radiation dose in excess of 25 rem TEDE.

Because the applicant has not selected a reactor design to be constructed on the proposed ESP site, the applicant used a PPE approach to demonstrate that it meets these requirements. A PPE is a set of plant design parameters that are expected to bound the characteristics of a reactor(s) that may be constructed at a site, and it serves as a surrogate for actual reactor design information. As discussed in RS-002 and in Chapter 1 of the SER (NUREG-1835), the staff considers the PPE approach to be an acceptable method for assessing site suitability. For the purposes of this analysis, the applicant proposed a fission product release from the PPE (ESP footprint) to the environment, and the staff reviewed the applicant's dose evaluation based on this release.

15.3 Technical Evaluation

The applicant evaluated the suitability of the site under the radiological consequence evaluation factors identified in 10 CFR 50.34(a)(1) using bounding reactor accident source terms and dose consequences as a set of PPE values based on three surrogate designs, as well as site-specific χ/Q values based on the ESP footprint. The following paragraphs describe the staff's review of each aspect of this evaluation.

15.3.1 Selection of DBAs

The applicant selected the DBAs listed in Section 15.1 of NUREG-1835 based on the proposed AP1000 reactor design, the certified ABWR design, and the proposed ESBWR design. The applicant indicated that it chose these three reactor designs because they have (or are based on) previously certified standard designs and have recognized bases for postulated accident analyses. The staff finds that the applicant selected DBAs that are consistent with the DBAs listed and analyzed in NUREG-0800 and RG 1.183. The applicant did not omit any applicable DBA identified in these guidance documents. Therefore, the staff finds that the applicant provided an acceptable DBA selection for evaluating the compliance of the proposed ESP site with the dose consequence evaluation factors specified in 10 CFR 50.34(a)(1). The applicant stated that, because of their greater potential for inherent safety, it expects the DBAs of the other reactors being considered for the proposed ESP site to be bounded by those DBAs analyzed in the proposed AP1000 and ESBWR and the certified ABWR DCDs. While the staff has not reviewed in detail designs other than the proposed AP1000 and certified ABWR, it believes that conclusions drawn regarding the site's acceptability based on the AP1000, ABWR, and ESBWR designs are likely to be valid for the other reactor designs that the applicant is considering. If a COL or CP application referencing any ESP that might be issued for the North Anna ESP site is filed, the applicant will confirm, and the staff will evaluate, whether the source term considered here bounds that of the design proposed in the COL or CP application.

15.3.2 Design-Specific (Postulated) χ/Q Values

The staff here discusses the χ/Qs postulated in the ABWR and AP1000 design certification proceedings because the applicant used these χ/Qs in its analysis as follows. To support its accident analyses based on the ABWR as a surrogate design, the applicant used the postulated χ/Q values in the certified ABWR DCD. In evaluating the AP1000, however, the applicant used those χ/Q values in the proposed (surrogate) AP1000 DCD that were under review by the staff at the time the North Anna ESP application was submitted. Westinghouse subsequently revised the χ/Q values in the AP1000 DCD. Consequently, the postulated χ/Q values and the calculated design-specific doses used in the North Anna ESP application may differ from those associated with a certified API000 DCD. Nonetheless, the staff determined that the PPE values for the postulated χ/Q values associated with the surrogate AP1000 design used by the applicant in its accident analyses are reasonable and, therefore, that they are adequate to demonstrate that a reactor with design characteristics similar to an AP1000 could be sited at the proposed ESP site. Section 15.4 of the SSAR lists the χ/Q values the applicant used for the version of the AP1000 and the certified ABWR that it considered. The staff compared the χ/Q's postulated for the ABWR and AP1000 to those measured for the North Anna ESP site as described below in Section 15.3.5.

Design-specific χ/Q values were not used in calculating the DBA doses for the ESBWR. Rather, the DBA doses were calculated directly from activity releases using site-specific χ/Q values. This subject is discussed in more detail in Section 15.3.5.

In Table 1.3-1 of the SSAR, the applicant also listed a set of design-specific postulated χ/Q values, some of which neither the applicant nor the staff had used in their radiological consequence evaluations. The staff finds that the χ/Q values in Table 1.3-1, with the exception of those used in the applicant's dose assessments in Chapter 15 of the SSAR, are not needed to assess the suitability of the proposed site.

15.3.3 Site-Specific χ/Q Values

The staff reviewed the applicant's site-specific χ/Q values and performed an independent evaluation of atmospheric dispersion in accordance with the guidance provided in Section 2.3.4 of RS-002. The χ/Q values indicate the atmospheric dilution capability. Smaller χ/Q values are associated with greater dilution capability, resulting in lower radiological doses. The radiological consequences are thus inversely proportional to the χ/Q values. The applicant provided the site-specific χ/Q values used in its radiological consequence analyses in Table 1.9-1 of the SSAR, and the staff discusses and evaluates its χ/Q values in Section 2.3.4 of NUREG-1835.

The applicant used the atmospheric dispersion computer code PAVAN, described in the 1982 report NUREG/CR-2858, "PAVAN: An Atmospheric Dispersion Program for Evaluating Design Basis Accidental Releases of Radioactive Materials from Nuclear Power Stations," to derive its site-specific χ/Q values. Section 2.3.4 of NUREG-1835 indicates that a copy of the input files used by the applicant to execute PAVAN can be found in the applicant's response to RAI 2.3.4-1. The staff describes the PAVAN code calculations for the North Anna site in more detail in Section 2.3.4 of NUREG-1835. The staff finds the χ/Q values to be acceptable, as described in Section 2.3.4 of NUREG-1835. The staff intends to include these site-specific χ/Qs in any ESP that the NRC may issue for the North Anna ESP site.

15.3.4 Source Term Evaluation

To evaluate the suitability of the site using the radiological consequence evaluation factors in 10 CFR 50.34(a)(1), the applicant provided a set of bounding reactor accident source terms as a set of PPE values based on (1) the surrogate AP1000, 4386-megawatt thermal (MWt) ABWR, including 2 percent ECCS evaluation margin, and the ESBWR design source terms (as explained below) and (2) the site-specific χ/Qs based on the ESP footprint. The source terms are expressed as the timing and release rate of fission products to the environment from the proposed ESP site. The dose consequences are then derived from the source terms using established methods.

The surrogate AP1000 source terms are based on the guidance provided in RG 1.183. The methodologies and assumptions used by Westinghouse, the AP1000 vendor, in its radiological consequence analyses are consistent with the guidance provided in RG 1.183. The resulting doses calculated for the surrogate AP1000 design using the postulated site parameters meet the dose consequence evaluation factors specified in 10 CFR 50.34(a)(1) (i.e., 25 rem TEDE).

The methodologies and assumptions used by GE, the ABWR vendor, in its radiological consequence analyses for the ABWR design are consistent with the guidance provided in RGs 1.3 and 1.25. The ABWR source terms are based on the guidance in TID-14844. As set forth in the ABWR DCD, the resulting doses for the ABWR reactor design using the postulated site parameters meet the dose consequence evaluation factors specified in 10 CFR 100.11, "Determination of Exclusion Area, Low Population Zone, and Population Center Distance," which are 300 rem to the thyroid and 25 rem to the whole body. While the requirements of 10 CFR 100.11 are not applicable to an ESP, the staff notes that the Commission, in promulgating the final rule in Appendix A, "Design Certification Rule for the U.S. Advanced Boiling Water Reactor," to 10 CFR Part 52, "Early Site Permits; Standard Design Certifications; and Combined Licenses for Nuclear Power Plants," stated the following:

> The Commission has determined that with regard to the revised design basis accident radiation dose acceptance criteria in 10 CFR 50.34, the ABWR design meets the new dose criteria based on the NRC staff's radiological consequence analyses, provided that the site parameters are not revised.

"Standard Design Certification for the U.S. Advanced Boiling Water Reactor Design," Final Rule, 62 Fed. Reg. 25800, 25819-820. Accordingly, the certified ABWR design, in conjunction with assumed site parameters, meets the dose consequence evaluation factors specified in 10 CFR 100.11, as well as those specified in 10 CFR 50.34(a)(1).

In its site-specific DBA radiological consequence analyses, the applicant scaled the ABWR source terms and the resulting doses from the power level, certified under Appendix A to 10 CFR Part 52, of 4005 MWt to 4386 MWt for its version of the ABWR. The applicant used a linear scaling method. Because the fission product release rate is directly proportional to the fission product inventory if mitigating processes remain the same, and because the fission product inventory is directly proportional to reactor power, the staff finds this scaling methodology to be acceptable for the purposes of this evaluation.

The ESBWR source terms are based on the guidance provided in RG 1.183. The methodologies and assumptions used by GE, the ESBWR vendor, in its radiological consequence analyses are consistent with the guidance provided in RG 1.183. Dominion increased the activity levels for ESBWR DBA analyses by 25 percent to accommodate uncertainty in its design because the design certification review is not complete. The resulting doses calculated for the ESBWR design using the North Anna ESP site-specific χ/Qs meet the dose consequence evaluation factors specified in 10 CFR 50.34(a)(1) (i.e., 25 rem TEDE).

15.3.5 Radiological Consequence Evaluations

In determining the potential radiological consequences resulting from DBAs for the ABWR and the AP1000 at the proposed site, the applicant used the site-specific χ/Q values in conjunction with the DBA radiological consequences and the postulated χ/Q values provided in the certified ABWR DCD and the surrogate AP1000 DCD. The certified ABWR and the proposed AP1000 designs met the radiological consequence evaluation factors identified in 10 CFR 50.34 (a)(1) with their postulated χ/Q values.

The applicant used the ratios of the site-specific χ/Q values to those postulated in the ABWR DCD and AP1000 DCD to determine and demonstrate that the radiological consequences at the proposed site meet the requirements of 10 CFR 50.34, "Contents of Applications; Technical Information." The estimated site-specific χ/Q values for the proposed site are lower than those postulated in the ABWR DCD and AP1000 DCD, i.e., reflect greater dispersion than assumed in the design certification proceedings. The certified ABWR and the proposed AP1000 designs met the radiological consequence evaluation factors identified in 10 CFR 50.34(a)(1) with their postulated χ/Q values. Accordingly, the resulting DBA radiological consequences at the proposed site are lower than those provided in the ABWR DCD and AP1000 DCD and, therefore, meet the requirements of 10 CFR 50.34.

Review of the ESBWR design certification by the staff is not complete. Therefore, ESBWR DBA dose calculations were performed using the modified source terms (125 percent of proposed DCD values) and site-specific χ/Q values rather than using doses from the design document and the ratios of site-specific χ/Q values to design χ/Q values. The applicant used PAVAN to derive its site-specific χ/Q values. Using these χ/Q values, the proposed ESBWR design meets the radiological consequence evaluation factors identified in 10 CFR 50.34(a)(1).

The staff has verified the design-specific source terms the applicant provided and finds them to be consistent with those evaluated (or being evaluated) by the staff as part of the design certification reviews. Further, the staff finds that the references provided by the applicant and the methodology it used to determine the timing and release rate of fission products to the environment (and consequent radiological consequences) from the proposed ESP site are acceptable. Therefore, the staff finds the source terms from the PPE (ESP footprint) themselves to be reasonable and acceptable. The staff intends to include the site-specific x/Q values listed as site characteristics in Appendix A in any ESP that the NRC might issue for the North Anna site.

Table 15.3-1 identifies the following site χ/Q values as appropriate for inclusion in any ESP that the staff might issue for the North Anna ESP site.

Table 15.3-1 Staff's Proposed Site-Specific χ/Q Values

Location and Time Interval	χ/Q Value
0 to 2 hour EAB	2.26E-4 s/m^3
0 to 8 hour LPZ	2.05E-5 s/m^3
8 to 24 hour LPZ	1.36E-5 s/m^3
1 to 4 day LPZ	5.58E-6 s/m^3
4 to 30 day LPZ	1.55E-6 s/m^3

RS-002 calls for the staff to perform a confirmatory radiological consequence calculation. The design-related inputs to the applicant's dose calculation for the AP1000 and ABWR designs were directly extracted from design documentation previously submitted to and reviewed by the staff in connection with design certification applications. Because the applicant simply used the ratio of the site-specific χ/Q values to the postulated design χ/Q values, the staff did not consider an independent calculation to be useful or necessary and, therefore, did not perform one. For the ESBWR, the staff performed confirmatory dose calculations using the isotopic release rates provided by the applicant and site-specific χ/Q values. The results of these calculations confirm that the dose calculations performed by the applicant are correct.

The staff believes that basing the radiological consequences of the DBAs at the proposed site on the AP1000, ABWR, and ESBWR designs is likely to be valid for the other reactor designs the applicant is considering. Whether the final reactor design selected by the applicant at the North Anna ESP site is in fact bounded by the acceptance made here would be subject to review during the staff's consideration of any COL or CP application. In accordance with 10 CFR 52.79(a)(1), at the COL stage, the staff will evaluate whether the design of the facility falls within the parameters specified in an ESP, if one is issued for the North Anna ESP site.

Based on the above evaluation of the applicant's analysis methodology and inputs to that analysis, the staff finds that the applicant's conclusion that the radiological consequences for the chosen surrogate designs comply with the radiological consequence evaluation factors of 10 CFR 50.34(a)(1) is correct.

Based on the ESBWR source term and χ/Qs evaluated above, the applicant calculated radiological consequences at the EAB and the LPZ boundary. The applicant performed these calculations using the methodologies specified in RG 1.183 which the staff has identified as acceptable for this purpose. The results obtained by the applicant are below the TEDE does specified in 10 CFR 50.34(a)(I). Accordingly, the staff finds that the applicant's conclusion that the radiological consequences for the ESBWR design complies with the radiological consequence evaluation factors of 10 CFR 50.34(a)(1) is correct.

15.4 Conclusions

As described above, the applicant submitted its radiological consequence analyses using the site-specific χ/Q values and PPE source term values and concluded that the proposed site meets the radiological consequence evaluation factors identified in 10 CFR 50.34(a)(1).

Based on the reasons given above, the staff finds that the applicant's PPE values for source terms included as inputs to the radiological consequence analyses are reasonable. Further, the staff finds that the applicant's site-specific χ/Q values and dose consequence evaluation methodology are acceptable. Therefore, the staff concludes that the proposed distances to the EAB and the LPZ outer boundary of the proposed ESP site, in conjunction with the fission product release rates to the environment provided by the applicant as PPE values, are adequate to provide reasonable assurance that the radiological consequences of the DBAs will be within the radiological consequence evaluation factors set forth in 10 CFR 50.34(a)(1) for the proposed ESP site. This conclusion is subject to confirmation at the COL or CP stage that the design of the facility specified by the COL or CP applicant falls within the values of site characteristics and plant parameters specified in any ESP that might issue for the North Anna ESP site.

The staff further concludes that (1) the applicant has demonstrated that the proposed ESP site is suitable for power reactors with source term characteristics bounded by those of the ABWR (at 4386 MWt), AP1000, and ESBWR without undue risk to the health and safety of the public and (2) the applicant has complied with the requirements of 10 CFR 52.17 and 10 CFR Part 100.

19. CONCLUSIONS

In accordance with Subpart A, "Early Site Permits," of Title 10, Part 52, "Early Site Permits, Standard Design Certifications, and Combined Licenses for Nuclear Power Plants," of the *Code of Federal Regulations* (10 CFR Part 52), the staff of the U.S. Nuclear Regulatory Commission (NRC) reviewed the site safety analysis report and emergency planning information included in the early site permit (ESP) application submitted by Dominion Nuclear North Anna, LLC, for the North Anna ESP site. On the basis of its evaluation and independent analyses as discussed in this supplement and NRC technical report NUREG-1835, "Safety Evaluation Report for an Early Site Permit (ESP) at the North Anna ESP Site," the staff concludes that the North Anna ESP site characteristics comply with the requirements of 10 CFR Part 100, "Reactor Site Criteria," with the limitations and conditions proposed by the staff in this supplement and NRC technical report NUREG-1835 for inclusion in any ESP that might be issued. Further, for the reasons set forth in this supplement and NRC technical report NUREG-1835, the staff concludes that, taking into consideration the site criteria contained in 10 CFR Part 100, a reactor, or reactors, having characteristics that fall within the parameters for the site, and which meets the terms and conditions proposed by the staff in this supplement and NRC technical report NUREG-1835, can be constructed and operated without undue risk to the health and safety of the public. For the same reasons, the staff also concludes that issuance of the requested ESP will not be inimical to the common defense and security or to the health and safety of the public. If issued, the North Anna ESP may be referenced in an application to construct or to construct and operate a nuclear power reactor, or reactors, with a total generating capacity of up to 9000 megawatts (thermal) at the ESP site, subject to the terms and conditions of the permit.

APPENDIX A

PERMIT CONDITIONS, COL ACTION ITEMS, SITE CHARACTERISTICS, AND BOUNDING PARAMETERS

A.1 Permit Conditions

Permit Condition: The Commission's regulation in 10 CFR § 52.24 authorizes the inclusion of limitations and conditions in an ESP. A permit condition is not needed when an existing NRC regulation requires a future regulatory review of a matter to ensure adequate safety during design, construction, or inspection activities for a new plant. The staff is proposing that the Commission include eight permit conditions, which are set forth below, to control various safety matters.

Permit Condition No.	SER Section	Description
		2.1 - Introduction
1	2.1.2	The NRC staff proposes to include a condition in any ESP that might be issued in connection with this application to govern exclusion area control. This permit condition would require that approvals called for by State law for, among other matters, agreements providing for shared control of the North Anna ESP exclusion area, be obtained and the agreements executed before construction of a nuclear power plant begins under a construction permit or COL referencing the ESP.
2	2.1.2	The NRC staff proposes to include a condition in any ESP that might be issued in connection with this application requiring that the ESP holder obtain the right to implement the site redress plan before initiating any activities authorized by 10 CFR 52.25.

Permit Condition No.	SER Section	Description
		2.4 - Hydrology
3	2.4.1	The NRC staff proposes to include a condition in any ESP that might be issued in connection with this application requiring that an applicant referencing such an ESP in an application for a fourth proposed unit use a dry cooling tower system during normal operation.
4	2.4.13	The NRC staff proposes to include a condition in any ESP that might be issued in connection with this application requiring that an applicant referencing such an ESP design any new unit's radwaste systems with features to preclude any and all accidental releases of radionuclides into any potential liquid pathway.
		2.5 - Geology, Seismology, and Geotechnical Engineering
5	2.5.1	The NRC staff proposes to include a condition in any ESP that might be issued in connection with this application requiring that the ESP holder and/or an applicant referencing such an ESP replace weathered or fractured rock at the foundation level with lean concrete before initiation of foundation construction.
6	2.5.1	The NRC staff proposes to include a condition in any ESP that might be issued in connection with this application prohibiting the ESP holder or an applicant referencing such an ESP from using an engineered fill with high compressibility and low maximum density, such as saprolite.
7	2.5.4	The NRC staff proposes to include a condition in any ESP that might be issued in connection with this application requiring that the ESP holder and/or an applicant referencing such an ESP perform geologic mapping of future excavations for safety-related structures, evaluate any unforeseen geologic features that are encountered, and notify the NRC no later than 30 days before any excavations for safety-related structures are open for NRC's examination and evaluation.
8	2.5.4	The NRC staff proposes to include a condition in any ESP that might be issued in connection with this application requiring that the ESP holder and/or an applicant referencing such an ESP improve Zone II saprolitic soils to reduce any liquefaction potential if safety-related structures are to be founded on them.

A.2 COL Action Items

COL Action Items: The combined license (COL) action items set forth in the SER and incorporated herein identify certain matters that shall be addressed in the final safety analysis report (FSAR) by an applicant who submits an application referencing the North Anna ESP. These items constitute information requirements but do not form the only acceptable set of information in the FSAR. An applicant may depart from or omit these items, provided that the departure or omission is identified and justified in the FSAR. In addition, these items do not relieve an applicant from any requirement in 10 CFR Parts 50 and 52 that govern the application. After issuance of a construction permit (CP) or COL, these items are not controlled by NRC requirements unless such items are restated in the preliminary safety analysis report or FSAR, respectively.

The staff identified the following COL action items with respect to individual site characteristics in order to ensure that particular significant issues are tracked and considered during the review of a later application referencing any ESP that might be issued for the North Anna ESP site.

Action Item No.	SER Section	Subject To Be Addressed	Reason for Deferral
		2.1 - Introduction	
2.1-1	2.1.1	A COL or CP applicant should provide latitude, longitude, and Universal Transverse Mercator coordinates for new units.	Exact unit locations not known at ESP stage.
2.1-2	2.1.2	A COL or CP applicant should make arrangements with the appropriate local, State, Federal, or other public agencies to provide for control of the portions of Lake Anna and the WHTF that are within the exclusion area.	Such arrangements not required at ESP stage.
		2.2 - Nearby Industrial, Transportation, and Military Facilities	
2.2-1	2.2.2	A COL or CP applicant should perform an evaluation of industrial hazards, if any, associated with this site.	No hazard present, but zoning could allow them during ESP term.
2.2-2	2.2.3	A COL or CP applicant should assess design-specific interactions between the existing and new units and, if necessary, propose measures to account for such interactions..	New unit design and specific location not known at ESP stage

Action Item No.	SER Section	Subject To Be Addressed	Reason for Deferral
		2.3 - Meteorology	
2.3-1	2.3.2	A COL or CP applicant should consider as part of detailed engineering the potential impact on the design or operation of the proposed unit(s) of any cooling-tower-induced local increase in (1) ambient air temperature, (2) ambient air moisture content, or (3) moisture and salt deposition.	Cooling tower location and design not known at ESP stage.
2.3-2	2.3.4	A COL or CP applicant should assess dispersion of airborne radioactive materials to the control room.	Control room location and design not known at ESP stage.
2.3-3	2.3.5	A COL or CP application should verify specific release point characteristics and specific locations of receptors of interest used to generate the long-term (routine release) atmospheric dispersion site characteristics.	Exact release points and receptor locations not known at ESP stage.
		2.4 - Hydrology	
2.4-1	2.4.1	A COL or CP application should provide the NRC for review the layout of intake and discharge tunnels and the construction techniques to be used before commencement of construction activities.	The feasibility of the use of the existing discharge tunnel from the abandoned units is not known at the ESP stage.
2.4-2	2.4.1	A COL or CP applicant should develop a plant shutdown protocol for proposed Unit 3 when water surface elevation in Lake Anna falls to 242 ft MSL	Future uses and therefore low-level frequency not known at ESP stage. Water surface elevation of 73.8 m (242 ft) MSL is the applicant-proposed shutdown level for the new units.
2.4-3	2.4.1	Withdrawn	Withdrawn
2.4-4	2.4.2	A COL or CP applicant should show that the ESP site is graded such that any flooding caused by local intense precipitation will be discharged to Lake Anna even in the event that any and all active drainage systems may be blocked and unable to function.	Detailed design of the plants, including the site grade are beyond the scope of an ESP review.

Action Item No.	SER Section	Subject To Be Addressed	Reason for Deferral
2.4-5	2.4.2	A COL or CP applicant should show that all safety-related structures are located at elevations above the maximum water surface elevation produced by local intense precipitation, or that adequate flood protection measures are in place to ensure their safety.	Certain locations within the ESP site area can be at the flood elevation of the site in response to local intense precipitation. It is not feasible to determine flooding protection needs at the ESP stage in response to local intense precipitation because final site grade and drainage patterns are not yet known.
2.4-6	2.4.4	A COL or CP applicant should demonstrate that the UHS reservoirs are designed so as to satisfy the NRC's regulations.	Detailed engineering design of underground UHS reservoirs, should they be needed, to preclude uplift due to buoyancy is not within the scope of ESP review.
2.4-7	2.4.4	A COL or CP applicant should demonstrate that the UHS storage basins provide storage sufficient to meet 30-day emergency cooling water needs accounting for any and all losses including but not limited to seepage, evaporation, and icing for the selected plants, if the selected plant designs includes a UHS. Programmatic provisions should be provided for plant shutdown when the liquid water volume in the UHS storage basin is inadequate.	Detailed engineering design of underground UHS reservoirs, should they be needed, to ensure adequate capacity is not within the scope of ESP review.
2.4-8	2.4.8	A COL or CP applicant should address whether Lake Anna or the WHTF will be used for safety-related water withdrawals.	The ESP water budget analysis relies on independent UHS reservoirs only, but need for a UHS is not known at the ESP stage.
2.4-9	2.4.10	A COL or CP applicant should adequately address the issue of slope embankment protection during design of the intake structure.	Safety of intake structure from slope embankment failure is a part of intake structure design, which is beyond the scope of an ESP review.

Action Item No.	SER Section	Subject To Be Addressed	Reason for Deferral
2.4-10	2.4.11	A COL or CP applicant should identify the most restrictive cooling water needs to account for the frequency of low-flow conditions and related minimum water elevation in Lake Anna and propose corresponding actions.	Technical specifications for safe shutdown of the plant due to low water conditions are based on consideration of the details of the design of the normal cooling water heat sink that are not available at the ESP stage.
2.5 - Geology, Seismology, and Geotechnical Engineering			
2.5-1	2.5.1	A COL or CP applicant should perform additional borings to identify any weathered or fractured rock beneath the new foundations.	Exact unit locations not known at ESP stage.
2.5-2	2.5.4	A COL or CP applicant should submit plot plans and the profiles of all seismic Category I facilities for comparison with the subsurface profile and material properties.	Exact unit locations and design not known at ESP stage.
2.5-3	2.5.4	An ESP holder and/or a COL or CP applicant should submit excavation and backfill plans for NRC review.	Exact unit locations and design not known at ESP stage.
2.5-4	2.5.4	A COL or CP applicant should assess groundwater conditions as they affect foundation stability or detailed dewatering plans.	Exact unit locations and design not known at ESP stage.
2.5-5	2.5.4	A COL or CP applicant should perform additional soil column amplification /attenuation analyses.	Exact unit locations not known at ESP stage.
2.5-6	2.5.4	A COL or CP applicant should provide analysis of the stability of all planned safety-related facilities, including bearing capacity, rebound, settlement, and differential settlements under deadloads of fills and plant facilities, as well as lateral loading conditions.	Exact unit locations and design not known at ESP stage.
2.5-7	2.5.4	A COL or CP applicant should provide design-related criteria pertinent to structural design.	Exact unit locations and design not known at ESP stage.

Action Item No.	SER Section	Subject To Be Addressed	Reason for Deferral
2.5-8	2.5.4	A COL or CP applicant should provide specific plans for each proposed ground improvements technique it plans to employ so that the staff may determine whether the chosen techniques will ensure that Zone IIA saprolitic soils will be able to support safety-related foundations.	Exact unit locations and design not known at ESP stage.
2.5-9	2.5-4	A COL or CP applicant should determine the average shear-wave velocity of the material underlying the foundation for the reactor containment and verify that it is equal to or exceeds that of the chosen design.	Site average shear-wave velocity of the Zone III-IV bedrock slightly less than design value provided at ESP stage.
2.5-10	2.5.5	A COL or CP applicant should conduct a more detailed dynamic analysis of the stability of the existing slope and any new slopes using the safe-shutdown earthquake (SSE) ground motion.	Locations of safety-related structures relative to the existing or new slopes not known at ESP stage.
2.5-11	2.5.5	A COL or CP applicant should provide plot plans and cross sections/profiles of all safety-related slopes, and specify the measures that it will take to ensure the safety of slopes and any structures located adjacent to the slopes.	Locations of safety-related structures relative to the existing or new slopes not known at ESP stage.
11.1 - Radiological Effluent Release Dose Consequences from Normal Operations			
11.1-1	11.1.4	A COL or CP applicant should verify that the calculated gaseous and liquid effluent concentrations and radiological doses to members of the public from radioactive gaseous and liquid effluents for any facility to be built on the North Anna site are bounded by the radiological doses and gaseous and liquid effluent concentrations included in the ESP application and reviewed by the NRC. The COL applicant should also include in the radwaste (gaseous and liquid effluents) system all items of reasonably demonstrated technology that affect reductions in population dose to maintain doses as low as reasonably achievable (ALARA) in accordance with 10 CFR Part 50, Appendix I, II.D.	Specific details of how the new facility will control, monitor, and maintain radioactive gaseous and liquid effluents not known at ESP stage.

A-8

13.6 - Industrial Security

13.6-1	13.6	A COL or CP applicant should provide specific designs for protected area barriers.	Exact locations and design of barriers not known at ESP stage.

A.3 Site Characteristics

Site Characteristics: Based on site investigation, exploration, analysis and testing, the applicant initially proposes a set of site characteristics. These site characteristics are specific physical attributes of the site, whether natural or man-made. Site characteristics, if reviewed and approved by the staff, are specified in the ESP. The staff proposes to include the following site characteristics in any ESP that might be issued for the North Anna site.

Site Characteristic	Value	Definition
	2.1 - Introduction	
Exclusion Area Boundary	The perimeter of a 5000 ft radius circle from the center of the abandoned Unit 3 containment	The area surrounding the reactor, in which the reactor licensee has the authority to determine all activities including exclusion or removal of personnel and property from the area
Low Population Zone	6 mile radius circle centered at the Unit 1 containment building	The area immediately surrounding the exclusion area which contains residents
Population Center Distance	8 miles	The minimum allowable distance from the reactor to the nearest boundary of a densely populated center containing more than about 25,000 residents

Site Characteristic		Value	Definition
2.3 - Meteorology			
Ambient Air Temperature and Humidity			
Maximum Dry-Bulb Temperature	2% annual exceedance	90 °F with 75 °F concurrent wet-bulb	The ambient dry-bulb temperature (and coincident wet-bulb temperature) that will be exceeded 2 percent of the time annually
	0.4% annual exceedance	95 °F with 77 °F concurrent wet-bulb	The ambient dry-bulb temperature (and coincident wet-bulb temperature) that will be exceeded 0.4 percent of the time annually
	100-year return period	109 °F	The ambient dry-bulb temperature that has a 1 percent annual probability of being exceeded (100-year mean recurrence interval)
Minimum Dry-Bulb Temperature	99% annual exceedance	18 °F	The ambient dry-bulb temperature below which dry-bulb temperatures will fall 1 percent of the time annually
	99.6% annual exceedance	14 °F	The ambient dry-bulb temperature below which dry-bulb temperature will fall 0.4 percentof the time annually
	100-year return period	-19 °F	The ambient dry-bulb temperature for which a 1 percent annual probability of a lower dry-bulb temperature exists (100-year mean recurrence interval)

Site Characteristic		Value	Definition
Maximum Wet-Bulb Temperature	0.4% annual exceedance	79 °F	The ambient wet-bulb temperature that will be exceeded 0.4 percent of the time annually
	100-year return period	88 °F	The ambient wet-bulb temperature that has a 1 percent annual probability of being exceeded (100-year mean recurrence interval)
Basic Wind Speed			
3-s Gust		96 mi/hr	The 3-s gust wind speed at 33 ft above the ground that has a 1 percent annual probability of being exceeded (100-year mean recurrence interval)
Design-Basis Tornado			
Maximum Wind Speed		260 mi/hr	Maximum wind speed resulting from passage of a tornado having a probability of occurrence of 10^{-7} per year
Translational Speed		52 mi/hr	Translation component of the maximum tornado wind speed
Rotational Speed		208 mi/hr	Rotation component of the maximum tornado wind speed
Radius of Maximum Rotational Speed		150 ft	Distance from the center of the tornado at which the maximum rotational wind speed occurs

Site Characteristic	Value	Definition
Maximum Pressure Drop	1.5 lbf/in^2	Decrease in ambient pressure from normal atmospheric pressure resulting from passage of the tornado
Maximum Rate of Pressure Drop	0.76 lbf/in^2/s	Rate of pressure drop resulting from the passage of the tornado
Winter Precipitation		
100-Year Snowpack	30.5 lbf/ft^2	Weight of the 100-year return period snowpack (to be used in determining extreme winter precipitation loads for roofs)
48-Hour Probable Maximum Winter Precipitation	20.75 in. of water	Probable maximum precipitation during the winter months (to be used in conjunction with the 100-year snowpack in determining extreme winter precipitation loads for roofs)
Ultimate Heat Sink Ambient Air Temperature and Humidity		
Meteorological Conditions Resulting in the Minimum Water Cooling During Any 1 Day	78.9 °F wet-bulb temperature with coincident 87.7 °F dry-bulb temperature	Historic worst 1-day daily average of wet-bulb temperatures and coincident dry-bulb temperatures
Meotorological Conditions Resulting in the Minimum Water Cooling During Any Consecutive 5 days	77.6 °F wet-bulb temperature with coincident 80.9 °F dry-bulb temperature	Historic worst 5-day daily average of wet-bulb temperatures and coincident dry-bulb temperatures resulting in minimum water cooling
Meteorological Conditions Resulting in the Maximum Evaporation and Drift Loss During Any Consecutive 30 Days	76.3 °F wet-bulb temperature with coincident 79.5 °F dry-bulb temperature	Historic worst 30-day daily average of wet-bulb temperatures and coincident dry-bulb temperatures

Site Characteristic	Value	Definition
Meteorological Conditions Resulting in the Maximum Water Freezing in the UHS Water Storage Facility	322 °F degree-days below freezing	Historic maximum cumulative degree-days below freezing
Short-Term (Accident Release) Atmospheric Dispersion		
0–2 hr χ/Q Value @ EAB	2.26×10^{-4} s/m^3	The 0–2 hour atmospheric dispersion factor to be used to estimate dose consequences of accidental airborne releases at the EAB
0–8 hr χ/Q Value @ LPZ	2.05×10^{-5} s/m^3	The 0–8 hour atmospheric dispersion factor to be used to estimate dose consequences of accidental airborne releases at the LPZ
8–24 hr χ/Q Value @ LPZ	1.36×10^{-5} s/m^3	The 8–24 hour atmospheric dispersion factor to be used to estimate dose consequences of accidental airborne releases at the LPZ
1–4 day χ/Q Value @ LPZ	5.58×10^{-6} s/m^3	The 1–4 day atmospheric dispersion factor to be used to estimate dose consequences of accidental airborne releases at the LPZ

Site Characteristic	Value	Definition
Long-Term (Routine Release) Atmospheric Dispersion		
Annual Average Undepleted/No Decay χ/Q Value @ EAB, east-southeast, 0.88 mile	3.7×10^{-6} s/m^3	The maximum annual average EAB undepleted/no decay χ/Q value for use in determining gaseous pathway doses to the maximally exposed individual
Annual Average Undepleted/2.26 Day Decay χ/Q Value @ EAB, east-southeast, 0.88 mile	3.7×10^{-6} s/m^3	The maximum annual average EAB undepleted/2.26 day decay χ/Q value for use in determining gaseous pathway doses to the maximally exposed individual
Annual Average Depleted/8.00 Day Decay χ/Q Value @ EAB, east-southeast, 0.88 mile	3.3×10^{-6} s/m^3	The maximum annual average EAB depleted/8.00 day decay χ/Q value for use in determining gaseous pathway doses to the maximally exposed individual
Annual Average D/Q Value @ EAB, east-southeast, 0.88 mile	1.2×10^{-8} 1/m^2	The maximum annual average EAB D/Q value for use in determining gaseous pathway doses to the maximally exposed individual
Annual Average Undepleted/No Decay χ/Q Value @ Nearest Resident, north-northeast, 0.96 mile	2.4×10^{-6} s/m^3	The maximum annual average resident undepleted/no decay χ/Q value for use in determining gaseous pathway doses to the maximally exposed individual

Site Characteristic	Value	Definition
Annual Average Undepleted/2.26 Day Decay χ/Q Value @ Nearest Resident, north-northeast, 0.96 mile	2.4×10^{-6} s/m^3	The maximum annual average resident undepleted/2.26 day decay χ/Q value for use in determining gaseous pathway doses to the maximally exposed individual
Annual Average Depleted/8.00 Day Decay χ/Q Value @ Nearest Resident, north-northeast, 0.96 mile	2.1×10^{-6} s/m^3	The maximum annual average resident depleted/8.00 day decay χ/Q value for use in determining gaseous pathway doses to the maximally exposed individual
Annual Average D/Q Value @ Nearest Resident, north-northeast, 0.96 mile	7.2×10^{-9} 1/m^2	The maximum annual average resident D/Q value for use in determining gaseous pathway doses to the maximally exposed individual
Annual Average Undepleted/No Decay χ/Q Value @ Nearest Meat Animal, southeast, 1.37 mile	1.4×10^{-6} s/m^3	The maximum annual average meat animal undepleted/no decay χ/Q value for use in determining gaseous pathway doses to the maximally exposed individual
Annual Average Undepleted/2.26 Day Decay χ/Q Value @ Nearest Meat Animal, southeast, 1.37 mile	1.4×10^{-6} s/m^3	The maximum annual average meat animal undepleted/2.26 day decay χ/Q value for use in determining gaseous pathway doses to the maximally exposed individual

A-16

Site Characteristic	Value	Definition
Annual Average Depleted/8.00 Day Decay χ/Q Value @ NearestMeat Animal, southeast, 1.37 mile	1.2×10^{-6} s/m^3	The maximum annual average meat animal depleted/8.00 day decay χ/Q value for use in determining gaseous pathway doses to the maximally exposed individual
Annual Average D/Q Value @ Nearest Meat Animal, southeast, 1.37 mile	3.1×10^{-9} 1/m^2	The maximum annual average meat animal D/Q value for use in determining gaseous pathway doses to the maximally exposed individual
Annual Average Undepleted/No Decay χ/Q Value @ Nearest Veg. Garden, northeast, 0.94 mile	2.0×10^{-6} s/m^3	The maximum annual average vegetable garden undepleted/no decay χ/Q value for use in determining gaseous pathway doses to the maximally exposed individual
Annual Average Undepleted/2.26 Day Decay χ/Q Value @ Nearest Veg. Garden, northeast, 0.94 mile	2.0×10^{-6} s/m^3	The maximum annual average vegetable garden undepleted/2.26 day decay χ/Q value for use in determining gaseous pathway doses to the maximally exposed individual
Annual Average Depleted/8.00 Day Decay χ/Q Value @ Nearest Veg. Garden, northeast, 0.94 mile	1.8×10^{-6} s/m^3	The maximum annual average vegetable garden depleted/8.00 day decay χ/Q value for use in determining gaseous pathway doses to the maximally exposed individual

Site Characteristic	Value	Definition
Annual Average D/Q Value @ Nearest Veg. Garden, northeast, 0.94 mile	6.0×10^{-9} $1/m^2$	The maximum annual average vegetable garden D/Q value for use in determining gaseous pathway doses to the maximally exposed individual
Annual Average Undepleted/No Decay X/Q Value; Annual Average Undepleted/2.26 Day Decay X/Q Value; and Annual Average Depleted/8.00 Day Decay X/Q Value @ Nearest Cow-Milk.	No value provided	The milk exposure pathway was not considered because there are no reported cows or goats used for milk production in the near vicinity of the site, within 5 miles. See discussion in ER Section 5.4.
Annual Average D/Q Value @ Nearest Cow-Milk.	No value provided	Same as above.

2.4 - Hydrology

Hydrology

Proposed Facility Boundaries	Appendix A, Figure 1 (FSER Figure 2.4.14-1) shows the proposed facility boundary using its corners numbered 1-8 and also lists the geographical coordinates of these points in Virginia State Plane Coordinate System using NAD 83 Datum. The coordinates are expressed in feet.	ESP site boundary map
Minimum Lake Water Level	242 ft MSL	Low water surface shutdown elevation for operation of NAPS Units 1 and 2, and of proposed Unit 3

Site Characteristic	Value	Definition
Maximum Elevation of Ground Water	82.3 m (270 ft) MSL or 1 ft below the free surface, whichever is higher	The maximum elevation of ground water at the ESP site
Flood Elevation	82.3 m (270 ft) MSL	Maximum flood level at the ESP site due to a PMF in Lake Anna's watershed, simultaneous failure of upstream storage reservoirs, and coincident wind-wave action.
Local Intense Precipitation	46.61 cm (18.35 in)/hour and 15.42 cm (6.07 in) in 5 minutes	Maximum potential rainfall at the immediate ESP site.
Frazil and Anchor Ice	The ESP site has the potential for formation of frazil and anchor ice.	Accumulated ice formation in a turbulent flow condition.
Maximum Ice Thickness	43.4 cm (17.1 in) thick	Ice sheet thickness at Lake Anna (based on maximum cumulative degree-days below freezing of 178.8 °C (321.8 °F))
Maximum Cumulative Degree-Days Below Freezing	178.8 °C (321.8 °F)	A measure of severity of winter weather conditions conducive to ice formation (computed using air temperature data from Piedmont Research Station)
Hydraulic Conductivity	1.0 m/d (3.4 ft/d)	Ground water flow rate per unit hydraulic gradient.

Site Characteristic	Value	Definition
Hydraulic Gradient	0.03 m/m (0.1 ft/ft)	Slope of ground water surface under unconfined conditions or slope of hydraulic pressure head under confined conditions.

A-20

Site Characteristic		Value	Definition
2.5 - Geology, Seismology, and Geotechnical Engineering			
Basic Geologic and Seismic Information			
Capable Tectonic Structures		------	No fault displacement potential within the investigative area
Vibratory Ground Motion			
Design Response Spectra		Appendix A, Figure 2 (FSER Figure 2.5.2-6)	Site Specific response spectra
Stability of Subsurface Materials and Foundations			
Zone III Weathered Rock (205ft - 298ft)	Minimum Bearing Capacity	16 ksf	Allowable load-bearing capacity of layer supporting plant structures
	Minimum Shear Wave Velocity	2000 ft/sec	Propagation of shear waves through foundation materials
Zone III - IV	Minimum Bearing Capacity	80 ksf	Allowable load-bearing capacity of layer supporting plant structures
	Minimum Shear Wave Velocity	3300 ft/sec	Propagation of shear waves through foundation materials
Zone IV Bedrock (188ft - 298ft)	Minimum Bearing Capacity	160 ksf	Allowable load-bearing capacity of layer supporting plant structures
	Minimum Shear Wave Velocity	6300 ft/sec	Propagation of shear waves through foundation materials

A.4 Bounding Parameters

Plant Parameter Envelope: A plant parameter envelope (PPE) sets forth postulated values of design parameters that provide design details to support the NRC staff's review of an ESP application. A controlling PPE value, or bounding parameter value, is one that necessarily depends on a site characteristic. As the PPE is intended to bound multiple reactor designs, the actual design selected in a combined license (COL) or construction permit (CP) application referencing an ESP would be reviewed to ensure that the design fits within the bounding parameter values. Otherwise, the COL or CP applicant would need to demonstrate that the design, given the site characteristics in the ESP, complies with the Commission's regulations. Should an applicant reference an ESP for a design that is not certified, the applicant would need to demonstrate that the design's characteristics fall within the bounding parameter values.

Bounding Parameters	Value	Definition
	2.4 - Hydrology	
Maximum Cooling Water Flow Rate - Unit 3	49.6 cfs	Maximum instantaneous withdrawal rate from the North Anna reservoir.
Minimum Site Grade	82.6 m (271 ft) MSL	Finished site grade

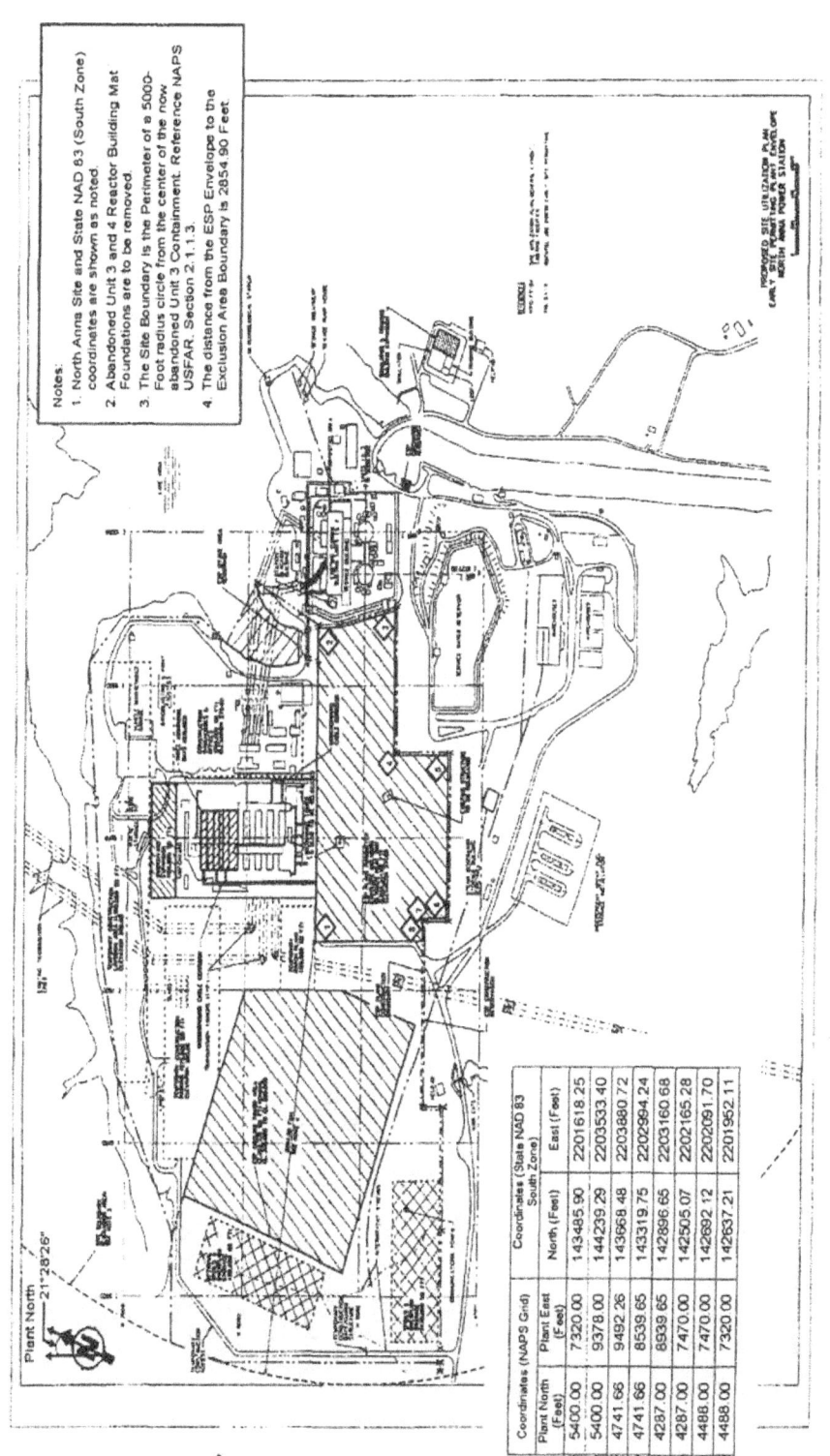

Figure 1 (Figure 2.4.14-1) The proposed facility boundary for the ESP site

A-23

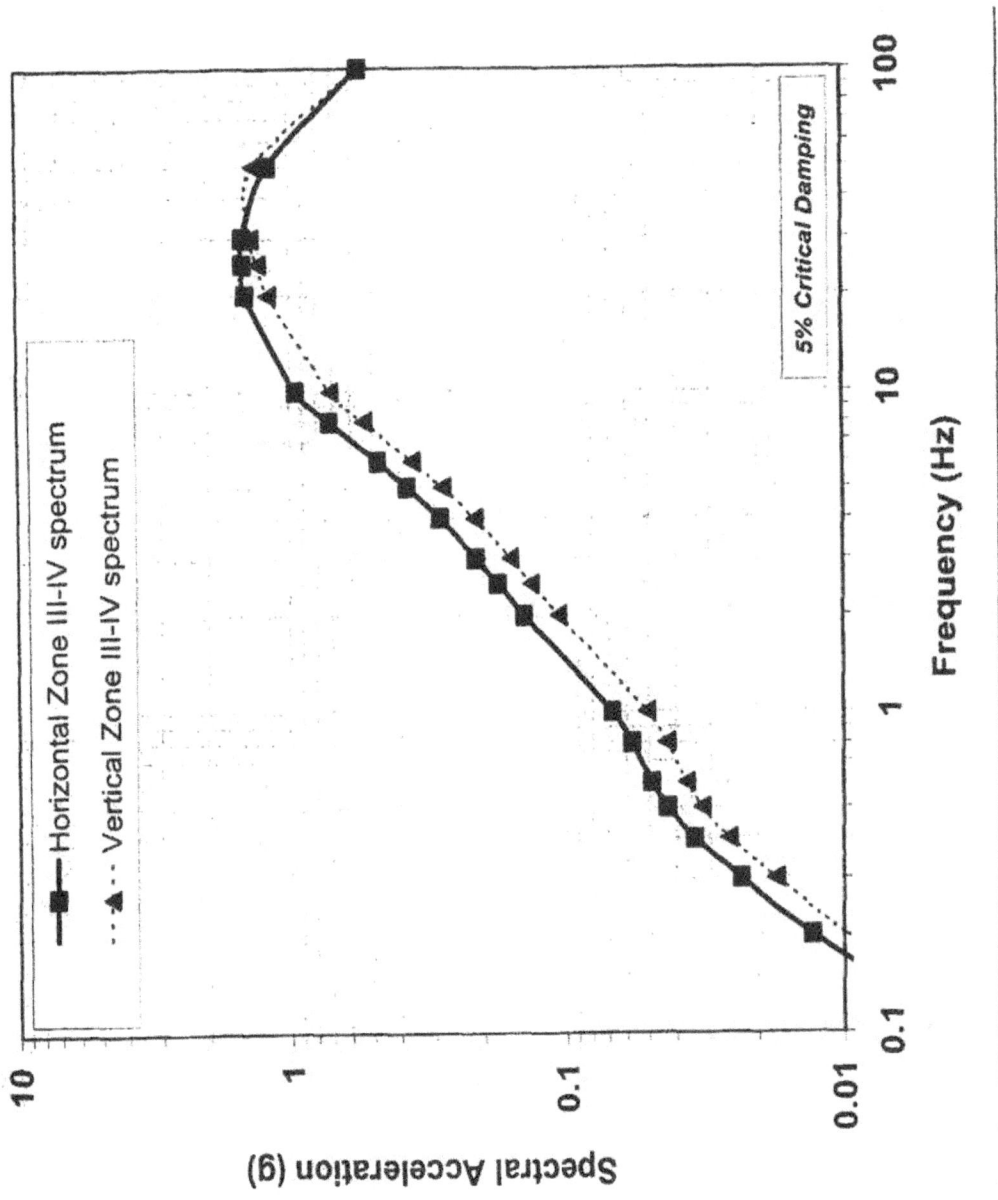

Figure 2 (Figure 2.5.2-6 (SSAR Figure 2.5-48A)) Selected Horizontal and Vertical Response Spectra for the Hypothetical Rock Outcrop Control Point SSE at the Top of Zone III-IV Material

A-24

APPENDIX B

CHRONOLOGY OF EARLY SITE PERMIT APPLICATION
FOR THE NORTH ANNA SITE

This appendix contains a chronological listing of routine licensing correspondence between the staff of the U.S. Nuclear Regulatory Commission (NRC) and Dominion Nuclear regarding the review of the North Anna early site permit application under Project No. 719 and Docket No. 52-008.

Revisions to the North Anna Early Site Permit Application

Rev.	Date	Accession Number
0	September 25, 2003	ML032731517
1˙	October 2, 2003	ML032731517
2	July 15, 2004	ML042010010
3	September 7, 2004	ML042590082
4	May 12, 2005	ML051450310
5	July 31, 2005	ML052150226
6	April 13, 2006	ML061180220
7	June 21, 2006	ML061870043
8	July 31, 2006	ML062140009
9	September 12, 2006	ML062580096

˙Revision 0 and Revision 1 of the application are contained in the same ADAMS package. Revision 1 of the application provides changes to Revision 0 to remove proprietary information from the application.

Document Date	Accession Number	Title/Description Includes Est. Page Count	Document Type	Author Affiliation(s)	Addressee Affiliation(s)	Docket Number
10/06/2005	ML052800406	2005/10/06-Email dated 10/6/2005 from Dominion to NRC, transmitting response requests for additional information submitted by NRC July 20, 2005. 8 Page(s)	E-Mail	Dominion Generation	NRC	05200008
10/06/2005	ML052790657	2005/10/06-North Anna Early Site Permit Application Response to Supplemental Request for Additional Information. 6 Page(s)	Letter	Dominion Nuclear North Anna, LLC	NRC/Docume nt Control Desk	05200008
10/24/2005	ML052980117	North Anna Early Site Permit Application Planned Revision to Unit 3 Cooling Water Approach. 4 Page(s)	Letter	Dominion Nuclear North Anna, LLC	NRC/Docume nt Control Desk	05200008
10/26/2005	ML053330014	2005/10/26-Email dated 10/26/2005 from Tony Banks, Dominion, to NRC Regarding the North Anna ESP Letter to VDEQ with Updated Cooling System Approach. 5 Page(s)	E-Mail	Dominion Generation	NRC	05200008

Document Date	Accession Number	Title/Description Includes Est. Page Count	Document Type	Author Affiliation(s)	Addressee Affiliation(s)	Docket Number
11/02/2005	ML053000566	Modification of the Cooling System for Dominion Nuclear North Anna ESP Site. 7 Page(s)	Letter	NRC/NRR/ADRA/DNRL	Dominion Resources Services, Inc	05200008
11/17/2005	ML053330016	2005/11/17-E-mail dated 11/17/2005 between Tony Banks, Dominion to NRC regarding Dominions North Anna Site ESP Wetlands Letter to Army Corps of Engineers. 14 Page(s)	E-Mail	Dominion Nuclear North Anna, LLC	NRC	05200008
11/22/2005	ML053260619	North Anna Early Site Permit (ESP), Application Submittal Schedule for ESP Application. 4 Page(s)	Letter	Dominion Nuclear North Anna, LLC	NRC/Docume nt Control Desk	05200008
11/22/2005	ML053330013	2005/11/22-Email dated 11/22/2005 from Margaret Bennett, Dominion, to NRC submitting North Anna application submittal schedule for ESP application supplement. 6 Page(s)	E-Mail	Dominion Generation	NRC	05200008

Document Date	Accession Number	Title/Description Includes Est. Page Count	Document Type	Author Affiliation(s)	Addressee Affiliation(s)	Docket Number
12/05/2005	ML053210054	Revision to the North Anna Early Site Permit (ESP) Issuance Schedule. 6 Page(s)	Letter	NRC/NRR/ADRA/DNRL	Dominion Resources Services, Inc, Innsbrook Technical Ctr	05200008
12/12/2005	ML053540012	2005/12/12-Summary of Telephone Conference with Dominion Nuclear North Anna, LLC to Discuss Revising the Early Site Permit (ESP) Application. 7 Page(s)	Meeting Summary	NRC/NRR/ADRO/DLR/REBA	Dominion Nuclear North Anna, LLC	05200008
01/13/2006	ML060200120	2006/01/13-E-Mail re: North Anna ESP Supplement. 329 Page(s)	E-Mail	Dominion Generation	NRC/NRR/ADRA/DNRL/NRBA	05200008
01/13/2006	ML060250396	2006/01/13-North Anna Early Site Permit Application Supplement to Address a Modified Approach to Unit 3 Cooling and to Ensure the Plant Parameter Envelope Remains Bounding. 327 Page(s)	Letter, Report, Miscellaneous	Dominion Nuclear North Anna, LLC	NRC/Document Control Desk	05200008

Document Date	Accession Number	Title/Description Includes Est. Page Count	Document Type	Author Affiliation(s)	Addressee Affiliation(s)	Docket Number
02/10/2006	ML060390208	2006/02/10-North Anna ESP Application Review Schedule. 10 Page(s)	Letter, Request for Additional Information (RAI)	NRC/NRR/ADRA /DNRL/NRBA	Dominion Resources, Inc	05200008
02/15/2006	ML060620148	2006/02/15-E-Mail re: Telecon Draft Talking Points. 14 Page(s)	E-Mail	NRC/NRR/ADRA /DNRL/NRBA	Dominion Generation	05200008
03/02/2006	ML060610065	2006/03/02-North Anna, Request for Additional Information, Results of Review of the Supplement to the ESP Application for the North Anna Site. 16 Page(s)	Letter, Request for Additional Information (RAI)	NRC/NRR/ADRA /DNRL/NRBA	Dominion Resources Services, Inc	05200008
03/02/2006	ML060790469	2006/03/02-E-Mail re: Dominion Participants on Recent Phone Calls. 3 Page(s)	E-Mail	Dominion Generation	NRC/NRR/AD RA/DNRL/NR BA	05200008
03/08/2006	ML060790473	2006/03/08-E-Mail re: Attendees for Friday Meeting at NRC. 2 Page(s)	E-Mail	Dominion Generation	NRC/NRR/AD RA/DNRL/NR BA	05200008

Document Date	Accession Number	Title/Description Includes Est. Page Count	Document Type	Author Affiliation(s)	Addressee Affiliation(s)	Docket Number
03/10/2006	ML060860087	2006/03/10-Attachment 4 - North Anna ESP Application 3/10/06 Dominion/NRC Meeting, Docket No. 52-008 - Meeting Handout. 24 Page(s)	Meeting Briefing Package/Hand outs	NRC/NRR/ADRA /DNRL/NRBA	Dominion Nuclear North Anna, LLC	05200008
03/10/2006	ML060860363	2006/03/10-Attachment 5 - NRC 3/2/06 Additional Information Needs and Discussion. 18 Page(s)	Meeting Briefing Package/Hand outs	NRC/NRR/ADRA /DNRL/NRBA	Dominion Nuclear North Anna, LLC	05200008
03/13/2006	ML060650396	2006/03/13-North Anna, Request for Additional Information, ESP Application for the North Anna Site. 5 Page(s)	Letter, Request for Additional Information (RAI)	NRC/NRR/ADRA /DNRL	Dominion Resources Services, Inc	05200008
03/16/2006	ML060810259	2006/03/16-E-Mail re: Fwd: Draft Meeting Summary. 30 Page(s)	E-Mail, Meeting Summary	NRC/NRR/ADRA /DNRL/NRBA	Dominion Generation	05200008
03/20/2006	ML060900316	2006/03/20-E-Mail re: Re: Fwd: Draft Meeting Summary. 2 Page(s)	E-Mail	Dominion Generation	NRC/NRR/AD RA/DNRL/NR BA	05200008

Document Date	Accession Number	Title/Description Includes Est. Page Count	Document Type	Author Affiliation(s)	Addressee Affiliation(s)	Docket Number
03/22/2006	ML060900297	2006/03/22-E-Mail re: North Anna Early Site Permit Application - March 10 Meeting Documents. 46 Page(s)	E-Mail, Slides and Viewgraphs	Dominion Generation	NRC/NRR/AD RA/DNRL/NR BA	05200008
03/23/2006	ML060900311	2006/03/23-E-Mail re: Invitation - NRC/GE Call on MACCS2 Inputs (Mar 23 04:00 PM EST in 2-SW-A). 3 Page(s)	E-Mail	Dominion Generation	NRC/NRR/AD RA/DNRL/NR BA	05200008
03/24/2006	ML060900312	2006/03/24-E-Mail re: One More Comment. 2 Page(s)	E-Mail	Dominion Generation	NRC/NRR/AD RA/DNRL/NR BA	05200008
04/01/2006	ML060950586	2006/04/01-E-Mail re: Re: Fwd: Draft Meeting Summary. 2 Page(s)	E-Mail	Dominion Generation	NRC/NRR/AD RA/DNRL/NR BA	05200008
04/03/2006	ML061040608	2006/04/03-North Anna Early Site Permit Application, Response to NRC Question 10.q - Water Budget Analysis Spreadsheets. 4 Page(s)	Letter	Dominion Nuclear North Anna, LLC	NRC/Docume nt Control Desk	05200008

Document Date	Accession Number	Title/Description Includes Est. Page Count	Document Type	Author Affiliation(s)	Addressee Affiliation(s)	Docket Number
04/03/2006	ML061040611	2006/04/03-North Anna Early Site Permit Application, Revised Cooling Analysis #24830-G-042, Attachment 2. 5 Page(s)	Spreadsheet File	Dominion Nuclear North Anna, LLC	NRC/Docume nt Control Desk	05200008
04/03/2006	ML061040612	2006/04/03-North Anna Early Site Permit Application, Revised Cooling Analysis #24830-G-042, Attachment 3. 5 Page(s)	Spreadsheet File	Dominion Nuclear North Anna, LLC	NRC/Docume nt Control Desk	05200008
04/05/2006	ML060960290	2006/04/05-E-Mail re: Fwd: 06-198 NRC Question 10.g - Water Budget Analysis Spreadsheets. 7 Page(s)	E-Mail, Letter	Dominion Generation	NRC/NRR, NRC/NRR/AD RA/DNRL/NR BA	05200008
04/05/2006	ML061670032	2006/04/05-Dominion North Anna Early Site Permit Application NRC Question 10.g - Water Budget Analysis Spreadsheets. 7 Page(s)	E-Mail, Environmental Impact Statement	Dominion Nuclear North Anna, LLC	NRC	05200008

Document Date	Accession Number	Title/Description Includes Est. Page Count	Document Type	Author Affiliation(s)	Addressee Affiliation(s)	Docket Number
04/11/2006	ML060860307	2006/04/11- 03/10/2006-ummary of Category 1 Meeting w/ Dominion Nuclear North Anna, LLC, Regarding the Supplement to the North Anna ESP Application. 8 Page(s)	Meeting Summary	NRC/NRR/ADRA /DNRL/NRBA	Dominion Nuclear North Anna, LLC	05200008
04/13/2006	ML061290145	2006/04/13-E-Mail re: North Anna ESP Application, Revision 6. 94 Page(s)	E-Mail	Dominion Generation	NRC/NRR/AD RA/DNRL/NR BA	05200008
04/13/2006	ML061040523	Email from J. Hegner Forwarding Cover Letter to Revision 6 to the North Anna ESP Application. 94 Page(s)	E-Mail	Dominion Generation	NRC	05200008
04/13/2006	ML061210194	04/13/2006-E-Mail re: North Anna ESP Application, Revision 6. 94 Page(s)	E-Mail, Letter	Dominion Generation	NRC/NRR/AD RA/DNRL	05200008
04/13/2006	ML061180220	2006/04/13-North Anna Early Site Permit Application Response to NRC Questions and Revision 6 to the Plant Application. 92 Page(s)	Letter, Report, Miscellaneous	Dominion Nuclear North Anna, LLC	NRC/Docume nt Control Desk	05200008, PROJ0719

Document Date	Accession Number	Title/Description Includes Est. Page Count	Document Type	Author Affiliation(s)	Addressee Affiliation(s)	Docket Number
04/30/2006	ML061180194	2006/04/30-North Anna Early Site Permit Application Revision 6, Cover through Page 2.5.4A-66. 532 Page(s)	Report, Miscellaneous	Dominion Nuclear North Anna, LLC	NRC	05200008, PROJ0719
04/30/2006	ML061180218	2006/04/30-North Anna Early Site Permit Application - Part 3 - Environmental Report Page 3-3-1 through Part 4 Page 4-1-9. 560 Page(s)	Environmental Report	Dominion Nuclear North Anna, LLC	NRC	05200008, PROJ0719
04/30/2006	ML061180206	2006/04/30-North Anna Early Site Permit Application - Figure 2.5-46 through Table 1, Appendix B. 155 Page(s)	Quality Assurance Program, Report, Miscellaneous, Updated Final Safety Analysis Report (UFSAR)	Dominion Nuclear North Anna, LLC	NRC	05200008, PROJ0719
04/30/2006	ML061180210	2006/04/30-North Anna Early Site Permit Application - Part 3 - Environmental Report Table of Contents through Figure 2.4-6. 110 Page(s)	Environmental Report	Dominion Nuclear North Anna, LLC	NRC	05200008, PROJ0719

Document Date	Accession Number	Title/Description Includes Est. Page Count	Document Type	Author Affiliation(s)	Addressee Affiliation(s)	Docket Number
04/30/2006	ML061180214	2006/04/30-North Anna Early Site Permit Application - Part 3 - Environmental Report Page 3-2-94 through Page 3-2-263. 170 Page(s)	Environmental Report	Dominion Nuclear North Anna, LLC	NRC	05200008, PROJ0719
04/30/2006	ML061180205	2006/04/30-North Anna Early Site Permit Application - Figure 2.5-15 through Figure 2.5-45. 32 Page(s)	Map, Report, Miscellaneous	Dominion Nuclear North Anna, LLC	NRC	05200008, PROJ0719
04/30/2006	ML061180203	2006/04/30-North Anna Early Site Permit Application - Page 2-2-329 through Figure 2.5-14. 97 Page(s)	Report, Miscellaneous	Dominion Nuclear North Anna, LLC	NRC	05200008, PROJ0719
05/01/2006	ML061460201	2006/05/01-E-Mail re: Fwd: Site Tour/Audit. 4 Page(s)	E-Mail	NRC/NRR/ADRA /DNRL/NRBA	Dominion Generation	05200008

Document Date	Accession Number	Title/Description Includes Est. Page Count	Document Type	Author Affiliation(s)	Addressee Affiliation(s)	Docket Number
05/02/2006	ML061240029	2006/05/02-Transmittal Letter - Notice of Intent to Prepare a Supplement to the Draft Environmental Impact Statement for an Early Site Permit (ESP) At the North Anna ESP Site (TAC No. MC1128). 13 Page(s)	Letter	NRC/NRR/ADRA/DNRL	Dominion Resources Services, Inc	05200008
05/04/2006	ML061230005	2006/05/04-North Anna ESP Application Review Schedule. 7 Page(s)	Letter	NRC/NRR/ADRA/DNRL	Dominion Resources Services, Inc	05200008
05/05/2006	ML061460205	2006/05/05-E-Mail re: Re: Telecon 5/01/2006. 2 Page(s)	E-Mail	Dominion	NRC/NRR/AD RA/DNRL/NR BA	05200008
05/05/2006	ML061460206	2006/05/05-E-Mail re: North Anna ESP Application Revision 6 Review Schedule Letter dated 5/04/2006. 10 Page(s)	E-Mail	NRC/NRR/ADRA/DNRL/NRBA	Dominion, US Dept of Energy (DOE)	05200008
05/10/2006	ML061460212	2006/05/10- E-Mail re: North Anna ESP Rev. 06 RAI Letter dated 5/10/2006. 11 Page(s)	E-Mail	NRC/NRR/ADRA/DNRL/NRBA	Dominion Generation	05200008

Document Date	Accession Number	Title/Description Includes Est. Page Count	Document Type	Author Affiliation(s)	Addressee Affiliation(s)	Docket Number
05/10/2006	ML061290142	2006/05/10- Request For Additional Information (RAI) Regarding Revision 6 of the ESP Application for the North Anna Site. 9 Page(s)	Letter, Request for Additional Information (RAI)	NRC/NRR/ADRA /DNRL	Dominion Resources Services, Inc, Innsbrook Technical Ctr	05200008
05/12/2006	ML061320447	2006/05/12- 05/03-04/2006 Summary of Site Audit to Support Review of Early Site Permit Application for the North Anna Site. 11 Page(s)	Meeting Summary	NRC/NRR/ADRA /DNRL	Dominion Nuclear North Anna, LLC	05200008
05/17/2006	ML061310198	2006/05/17- 05/01/2006 Summary of Telephone Conference With Dominion Nuclear North Anna, LLC, Regarding North Anna ESP Review. 7 Page(s)	Note to File incl Telcon Record, Verbal Comm	NRC/NRR/ADRO /DLR/REBA	Dominion Nuclear North Anna, LLC	05200008
05/24/2006	ML061510131	2006/05/24- North Anna Early Site Permit Application, Response to NRC May 10, 2006 Request for Additional Information May 12, 2006 Site Audit Summary Report Comments, and NRC Site Audit Follow-up Questions. 66 Page(s)	Letter	Dominion Nuclear North Anna, LLC	NRC/Docume nt Control Desk	05200008

Document Date	Accession Number	Title/Description Includes Est. Page Count	Document Type	Author Affiliation(s)	Addressee Affiliation(s)	Docket Number
05/24/2006	ML061670026	2006/05/24-Dominion North Anna Early Site Permit Application Response to NRC May 10, 2006 Request for Additional Information, May 12, 2006 Site Audit Summary Report Comments and NRC Site Audit follow-up Questions. 70 Page(s)	E-Mail, Letter	Dominion Generation	Battelle Memorial Institute, Pacific Northwest National Lab, NRC	05200008
05/24/2006	ML061580522	2006/05/24-E-Mail re: Dominion May 24, 2006 Response to NRC RAIs re North Anna ESP Application. 69 Page(s)	E-Mail	Dominion Generation	Battelle Memorial Institute, Pacific Northwest National Lab, NRC/NRR/AD RA/DNRL/NR BA	05200008
05/30/2006	ML061530400	2006/05/30-Comment (12) submitted by Dominion Nuclear North Anna, LLC, Eugene S. Grecheck on Proposed Rules PR-1, 2, 10, 19, 20, 21, 25, 26, 50, 51, 52, 54, 55, 72,73,75,95, 140, 170 and 171 re Licenses, Certifications, and Approvals for Nuclear Plants. 23 Page(s)	Rulemaking-Comment	Dominion Nuclear North Anna, LLC	NRC/SECY/R AS	05200008

Document Date	Accession Number	Title/Description Includes Est. Page Count	Document Type	Author Affiliation(s)	Addressee Affiliation(s)	Docket Number
06/07/2006	ML061580174	2006/06/07-Summary of Telephone Conferences With Dominion Nuclear North Anna, LLC Regarding North Anna ESP Review. 10 Page(s)	Meeting Summary	NRC/NRR/ADRA /DNRL	Dominion Nuclear North Anna, LLC	05200008
06/08/2006	ML061730363	2006/06/08-E-Mail re: References - RAIs 4 and 6 of RAI Letter dated 5/10/2006. 2 Page(s)	E-Mail	NRC/NRR/ADRA /DNRL/NRBA	Dominion Generation	05200008
06/08/2006	ML061670058	North Anna Early Site Permit Responses to RAI About ESBWR LOCA Radiological Analysis in Reference to RAI 4 and 6 in 5/10/2006 Letter. 7 Page(s)	Calculation, E-Mail, Environmental Impact Statement	Dominion Generation	NRC	05200008
06/08/2006	ML061730360	2006/06/08-E-Mail re: Telecon Summary dated 6/7/2006-North Anna ESP Application Revision 06. 11 Page(s)	E-Mail	NRC/NRR/ADRA /DNRL/NRBA	Dominion Generation	05200008

Document Date	Accession Number	Title/Description Includes Est. Page Count	Document Type	Author Affiliation(s)	Addressee Affiliation(s)	Docket Number
06/08/2006	ML061670056	North Anna Early Site Permit Request for Additional Information Reference - RAIs 4 and 6 of RAI Letter dated 05/10/2006. 5 Page(s)	E-Mail, Environmental Impact Statement	Dominion Generation	NRC	05200008
06/08/2006	ML061670042	2006/06/08-North Anna Early Site Permit RAIs 4 and 6 of RAI Letter Dated 05/10/2006. 8 Page(s)	E-Mail	Dominion Generation	NRC	05200008
06/08/2006	ML061730364	2006/06/08-E-Mail re: Re: References--RAIs 4 and 6 of RAI Letter dated 5/10/2006. 14 Page(s)	E-Mail	Dominion Generation	NRC/NRR/AD RA/DNRL/NR BA	05200008
06/21/2006	ML061840360	2006/06/21-E-Mail re: 06-507 Response to NRC Questions and Revision 7 to the North Anna ESP Application. 22 Page(s)	E-Mail	Dominion Generation	NRC/NRR/AD RA/DNRL/NR BA	05200008
06/21/2006	ML061870043	2006/06/21-North Anna Early Site Permit Application, Response to NRC Questions and Revision 7 to the North Anna ESP Application. 18 Page(s)	Legal-Affidavit, Letter	Dominion Nuclear North Anna, LLC	NRC/Docume nt Control Desk	05200008

Document Date	Accession Number	Title/Description Includes Est. Page Count	Document Type	Author Affiliation(s)	Addressee Affiliation(s)	Docket Number
06/30/2006	ML061870047	2006/06/30-Transmittal of North Anna Early Site Permit Application, Revision 7. 2220 Page(s)	Environmental Report, Final Safety Analysis Report (FSAR), Letter, License-Application for Construction Permit DKT 50	Dominion Nuclear North Anna, LLC	NRC/Docume nt Control Desk, NRC/NRR	05200008
07/06/2006	ML061660030	2006/07/06-Ltr. - Notice of Availability of the Supplement to the Draft Environmental Impact Statement for an ESP at the North Anna ESP Site. 13 Page(s)	Letter	NRC/NRR/ADRA /DNRL	Dominion Nuclear North Anna, LLC	05200008
07/18/2006	ML061990240	2006/07/18- Summary of Telephone Conference and the Site Audit with Dominion Nuclear North Anna LLC Regarding the North Anna ESP Review. 12 Page(s)	Letter, Meeting Summary	NRC/NRR/ADRA /DNRL	Dominion Nuclear North Anna, LLC	05200008

B-17

Document Date	Accession Number	Title/Description Includes Est. Page Count	Document Type	Author Affiliation(s)	Addressee Affiliation(s)	Docket Number
07/18/2006	ML062230231	07/18/2006-E-Mail re: Telecon Summary dated 7/18/2006-North Anna ESP Application Revision 07. 14 Page(s)	E-Mail, Meeting Summary	NRC/NRR/ADRA /DNRL/NESB	Dominion Resources Services, Inc	05200008
07/31/2006	ML062350100	07/31/2006-E-Mail re: Fw: 06-631 Response to NRC Questions and Revision 8 to the North Anna ESP Application (Letter and Enclosures 1 and 2 Only). 27 Page(s)	E-Mail, Letter	Dominion Generation	NRC, NRC/NRR/AD RA/DNRL/NE SB	05200008
07/31/2006	ML062140010	2006/07/31-North Anna Early Site Permit Application Response to NRC Questions and Revision 8 to North Anna ESP Application. 25 Page(s)	Letter	Dominion Nuclear North Anna, LLC	NRC/Docume nt Control Desk	05200008
08/07/2006	ML062280472	Comment (W2) of Eric Cantor, on Behalf of the Dominion Power, Supporting Dominion's Early Site Permit (ESP) Application for the North Anna Power Station Site. 1 Page(s)	General FR Notice Comment Letter	Dominion, US Congress, US HR (House of Representatives)	NRC/ADM/DA S/RDB	05200008

Document Date	Accession Number	Title/Description Includes Est. Page Count	Document Type	Author Affiliation(s)	Addressee Affiliation(s)	Docket Number
08/15/2006	ML062350256	Comment (W4) of William M. Murphey, on Behalf of Lake Anna Civic Association, on North Anna Early Site Permit. Listed Recommendations Should be Implemented. 4 Page(s)	General FR Notice Comment Letter	Lake Anna Civic Association (LACA)	Dominion Resources, Inc, NRC, State of VA, Dept of Environmental Quality	05200008
08/15/2006	ML062210405	Letter to D. Christian, Dominion re: Supplement 1 to Final Safety Evaluation Report for the North Anna Early Site Permit Application. 7 Page(s)	Final Safety Evaluation Report (FSER), Letter	NRC/NRR/ADRA /DNRL	Dominion Resources Services, Inc, Innsbrook Technical Ctr	05200008
09/06/2006	ML062560013	09/06/2006-E-Mail re: Fw: North Anna ESP Site: Revision to Wetlands Delineation 28 Page(s)	E-Mail	Dominion Generation	Battelle Memorial Institute, Pacific Northwest National Lab, NRC/NRR/AD RA/DNRL/NE PB	05200008
09/12/2006	ML062580099	Transmittal of Rev. 9 to North Anna Early Site Permit Application. 11 Page(s)	Letter	Dominion Nuclear North Anna, LLC	NRC/Docume nt Control Desk	05200008

Document Date	Accession Number	Title/Description Includes Est. Page Count	Document Type	Author Affiliation(s)	Addressee Affiliation(s)	Docket Number
09/12/2006	ML062560365	North Anna Early Site Permit Application Comments on NUREG-1811, Supplement 1 Draft Environmental Impact Statement for an Early Site Permit at North Anna ESP Site. 28 Page(s)	Letter	Dominion Nuclear North Anna, LLC	NRC/Docume nt Control Desk	05200008
09/12/2006	ML062650159	09/12/2006-E-Mail re: SDEIS Comments From Dominion. 32 Page(s)	E-Mail	Dominion Generation	NRC/NRR/AD RA/DNRL/NE PB, NRC/NRR/AD RA/DNRL/NE SB	05200008
09/13/2006	ML062650163	09/13/2006-E-Mail re: Dominion North Anna ESPA, Rev. 9. 3 Page(s)	E-Mail	Dominion Generation	NRC/NRR/AD RA/DNRL/NE PB, NRC/NRR/AD RA/DNRL/NE SB	05200008

Document Date	Accession Number	Title/Description Includes Est. Page Count	Document Type	Author Affiliation(s)	Addressee Affiliation(s)	Docket Number
09/13/2006	ML062650165	09/13/2006-E-Mail re: 06-790 Revision 9 to the North Anna ESP Application. 14 Page(s)	E-Mail	Dominion Generation	NRC/NRR/AD RA/DNRL/NE PB, NRC/NRR/AD RA/DNRL/NE SB, Pacific National Lab	05200008
09/13/2006	ML062440233	08/15/2006–Summary of Public Meeting Held To Receive Comments on the Supplement to DEIS For the North Anna ESP Application w/Enclosure 1-List of Attendees. 12 Page(s)	Meeting Summary	NRC/NRR/ADRA /DNRL/NEPB	Dominion Nuclear North Anna, LLC	05200008
09/13/2006	ML062440240	Enclosure 2 - Corrected DEIS Transcript for North Anna Site. 180 Page(s)	Meeting Transcript	NRC/NRR/ADRA /DNRL/NEPB	Dominion Nuclear North Anna, LLC	05200008
09/26/2006	ML062620039	09/06 & 08/2006 Summary of Telephone Conference and Site Audit with Dominion Nuclear, North Anna, LLC Regarding the North Anna ESP Review. 7 Page(s)	Meeting Summary	NRC/NRR/ADRA /DNRL/NESB	Dominion Nuclear North Anna, LLC	05200008

APPENDIX C

REFERENCES

Dominion Nuclear North Anna, LLC

-----, October 24, 2005, Letter from Eugene S. Grecheck, Dominion Nuclear North Anna, LLC, to NRC, Subject: North Anna Early Site Permit Application: Planned Revision to Unit 3 Cooling Water Approach. (ADAMS Accession No. ML052980117)

-----, November 22, 2005, Letter from Eugene S. Grecheck, Dominion Nuclear North Anna, LLC, to NRC, Subject: North Anna Early Site Permit Application: Submittal Schedule for ESP Application Supplement. (ADAMS Accession No. ML053260619)

-----, January 13, 2006, Letter from Eugene S. Grecheck, Dominion Nuclear North Anna, LLC, to NRC, Subject: North Anna Early Site Permit Application: Supplement to Address A Modified Approach to Unit 3 Cooling and to Ensure the Plant Parameter Envelope Remains Bounding. (ADAMS Accession No. ML060250396)

-----, April 13, 2006, Letter from Eugene S. Grecheck, Dominion Nuclear North Anna, LLC, to NRC, Subject: North Anna Early Site Permit Application: Response to NRC Questions and Revision 6 to the North Anna ESP Application. (ADAMS Accession No. ML061180220)

-----, June 21, 2006, Letter from Eugene S. Grecheck, Dominion Nuclear North Anna, LLC, to NRC, Subject: North Anna Early Site Permit Application: Response to NRC Questions and Revision 7 to the North Anna ESP Application. (ADAMS Accession No. ML061870030)

-----, July 31, 2006, Letter from Eugene S. Grecheck, Dominion Nuclear North Anna, LLC, to NRC, Subject: North Anna Early Site Permit Application: Response to NRC Questions and Revision 8 to the North Anna ESP Application. (ADAMS Accession No. ML062140009)

-----, September 12, 2006, Letter from Eugene S. Grecheck, Dominion Nuclear North Anna, LLC, to NRC, Subject: North Anna Early Site Permit Application, Revision 9 to the North Anna Early Site Permit Application. (ADAMS Accession No. ML062580096)

U.S. Atomic Energy Commission (AEC)

-----, TID-14844, "Calculation of Distance Factors for Power and Test Reactor Sites," AEC: Washington, D.C., March 1962. (ADAMS Accession No. ML021750625)

U.S. Code of Federal Regulations

-----, *Title 10, Energy*, Part 20, "Standards for Protection Against Radiation."

-----, *Title 10, Energy*, Part 21, "Reporting of Defects and Noncompliance."

-----, *Title 10, Energy*, Part 40, "Environmental Radiation Protection Standards for Nuclear Power Operations."

-----, *Title 10, Energy*, Part 50, "Domestic Licensing of Production and Utilization Facilities."

-----, *Title 10, Energy*, Part 52, "Early Site Permits; Standard Design Certifications; and Combined Licenses for Nuclear Power Plants."

-----, *Title 10, Energy*, Part 100, "Reactor Site Criteria."

-----, *Title 40, Protection of Environment*, Part 190, "Environmental Radiation Protection Standards for Nuclear Power Operations."

U.S. Nuclear Regulatory Commission (NRC)

NUREG-Series Reports

-----, NUREG-0800, Revision 3, "Standard Review Plan for the Review of Safety Analysis Reports for Nuclear Power Plants," NRC: Washington, D.C., July 1981.

-----, NUREG-1835, "Safety Evaluation Report for an Early Site Permit (ESP) at the North Anna ESP Site," September 2005.

-----, NUREG/CR-2858, "PAVAN: An Atmospheric Dispersion Program for Evaluating Design Basis Accidental Releases of Radioactive Materials from Nuclear Power Stations," NRC: Washington, D.C., November 1982.

-----, NUREG/CR-4013, "LADTAP II - Technical Reference and User Guide," April 1986.

-----, NUREG/CR-4653, "GASPAR II - Technical Reference and User Guide," March 1987.

Regulatory Guides

-----, Regulatory Guide 1.3, Revision 2, "Assumptions Used for Evaluating the Potential Radiological Consequences of a Loss of Coolant Accident for Boiling Water Reactors," NRC: Washington, D.C., June 1974. (ADAMS Accession No. ML003739601)

-----, Regulatory Guide 1.25 (Safety Guide 25), "Assumptions Used for Evaluating the Potential Radiological Consequences of a Fuel Handling Accident in the Fuel Handling and Storage Facility for Boiling and Pressurized Water Reactors," NRC: Washington, D.C., March 1972. (ADAMS Accession No. ML003769781)

-----, Regulatory Guide 1.70, Revision 3, "Standard Format and Content of Safety Analysis Reports for Nuclear Power Plants - LWR Edition," NRC: Washington, D.C., November 1978. (ADAMS Accession Nos. ML003740072, ML003740108, & ML003740116).

-----, Regulatory Guide 1.109, Revision 1, "Calculation of Annual Doses to Man from Routine Releases of Reactor Effluents for the Purpose of Evaluating Compliance with 10 CFR Part 50, Appendix I," NRC: Washington, D.C., October 1977. (ADAMS Accession No. ML003740384)

-----, Regulatory Guide 1.111, Revision 1, "Methods for Estimating Atmospheric Transport and Dispersion of Gaseous Effluents in Routine Releases from Light-Water-Cooled Reactors," NRC: Washington, D.C., July 1977. (ADAMS Accession No. ML003740354)

-----, Regulatory Guide 1.145, Revision 1, "Atmospheric Dispersion Models for Potential Accident Consequence Assessments at Nuclear Power Plants," NRC: Washington, D.C., February 1983. (ADAMS Accession No. ML003740205)

-----, Regulatory Guide 1.183, "Alternative Radiological Source Terms for Evaluating Design Basis Accidents at Nuclear Power Reactors," NRC: Washington, D.C., July 2000. (ADAMS Accession No. ML003716792)

Other NRC Documents

-----, General Design Criterion (GDC) 2, "Design Bases for Protection Against Natural Phenomena."

-----, NRR Review Standard, RS-002, "Processing Applications for Early Site Permits," NRC: Washington, D.C., May 3, 2004. (ADAMS Accession No. ML040700236)

Other References

-----, Federal Guidance Report 11, "Limiting Values of Radionuclide Intake and Air Concentration and Dose Conversion Factors for Inhalation, Submersion, and Ingestion," 1988.

-----, Federal Guidance Report 12, "External Exposure to Radionuclides in Air, Water, and Soil," 1993.

APPENDIX D

PRINCIPAL CONTRIBUTORS

Name	Responsibility
Bagchi, Goutam	Hydrology
Tammara Seshagiri	Site Hazards
Harvey, Robert B.	Meteorology
Dehmel Jean-Claude ·	Normal Radiological Dose Analyses
Lee, Jay	Accident Analyses
Munson, Cliff	Geology and Seismology
Musico, Bruce	Emergency Planning
Prescott, Paul	Quality Assurance
Nitin Patel	Project Management
Segala, John	Project Management
Wunder George	Project Management
Tardiff, Albert	Security

Contractors	Technical Area
Pacific Northwest Laboratory	Hydrology, Meteorology, and Normal Radiological Dose Analyses

APPENDIX E
UNITED STATES
NUCLEAR REGULATORY COMMISSION
ADVISORY COMMITTEE ON REACTOR SAFEGUARDS
WASHINGTON, DC 20555 - 0001

October 13, 2006

MEMORANDUM TO: Luis A. Reyes
Executive Director for Operations
/RA/

FROM: John T. Larkins, Executive Director
Advisory Committee on Reactor Safeguards

SUBJECT: SUPPLEMENT 1 TO FINAL SAFETY EVALUATION REPORT FOR
NORTH ANNA EARLY SITE PERMIT (ESP) APPLICATION

During the 536th meeting of the Advisory Committee on Reactor Safeguards, October 4-6, 2006, the Committee considered the changes reflected in Revisions 6, 7, 8, and 9 of Dominion Nuclear North Anna LLC (Dominion) application for an early site permit (ESP). In its revised application, Dominion proposed: (1) to change the once-through cooling system planned for Unit 3 in previous versions of the safety site analysis report (SSAR) to a closed-cycle system; (2) to increase the power levels for Units 3 and 4 to match the designed maximum power (4500 MWt) of a General Electric Economic and Simple Boiling-Water Reactor (ESBWR), one of the reactor designs included in the plant parameter envelope; and (3) to reduce the bounding value for tritium activity release (associated with the ACR-700 design), to ensure that the tritium concentration in liquid effluent releases is less than both the 10 CFR Part 20 limit and the limit set in the EPA drinking water standards. By letter dated September 29, 2006, the staff transmitted Supplement 1 to its final Safety Evaluation Report (SER), which addresses Revisions 6 through 9 of the North Anna ESP application, to the ACRS for possible review.

The Committee decided that the proposed changes do not affect its previous conclusions and recommendations with regard to issuing the ESP, and that additional review of this document prior to issuance is not necessary.

References:

1. Memorandum dated September 29, 2006, from David B. Matthews, Director, Division of New Reactor Licensing, NRR, to John T. Larkins, Executive Director, ACRS, Subject: Transmittal of Supplement 1 to Final Safety Evaluation Report for North Anna Early Site Permit (ESP) Application.
2. U.S. Nuclear Regulatory Commission, Final Safety Evaluation Report, "Safety Evaluation Report for an Early Site Permit (ESP) at the North Anna ESP Site," dated September 2005 (NUREG-1835).

NRC FORM 335
(9-2004)
NRCMD 3.7

U.S. NUCLEAR REGULATORY COMMISSION

1. REPORT NUMBER
(Assigned by NRC, Add Vol., Supp., Rev., and Addendum Numbers, if any.)

NUREG-1835
Supplement 1

BIBLIOGRAPHIC DATA SHEET

(See instructions on the reverse)

2. TITLE AND SUBTITLE

Safety Evaluation Report for an
Early Site Permit (ESP) at the
North Anna ESP Site

3. DATE REPORT PUBLISHED

MONTH	YEAR
November	2006

4. FIN OR GRANT NUMBER

5. AUTHOR(S)

6. TYPE OF REPORT

Technical

7. PERIOD COVERED (Inclusive Dates)

8. PERFORMING ORGANIZATION - NAME AND ADDRESS (If NRC, provide Division, Office or Region, U.S. Nuclear Regulatory Commission, and mailing address; if contractor, provide name and mailing address.)

Division of New Reactor Licensing
Office of New Reactors
U.S. Nuclear Regulatory Commission
Washington, D.C. 20555-0001

9. SPONSORING ORGANIZATION - NAME AND ADDRESS (If NRC, type "Same as above"; if contractor, provide NRC Division, Office or Region, U.S. Nuclear Regulatory Commission, and mailing address.)

Same as above

10. SUPPLEMENTARY NOTES

Docket No. 52-008, Project No. 719

11. ABSTRACT (200 words or less)

NUREG-1835 documents the U.S. Nuclear Regulatory Commission (NRC) staff's technical review of the site safety analysis report and emergency planning information included in the early site permit (ESP) application submitted by Dominion Nuclear North Anna, LLC (Dominion or the applicant), for the North Anna ESP site. By letter dated September 25, 2003, Dominion submitted the ESP application for the North Anna ESP site in accordance with Subpart A, 'Early Site Permits," of Title 10, Part 52, 'Early Site Permits; Standard Design Certifications; and Combined Licenses for Nuclear Power Plants," of the Code of Federal Regulations. Subsequently, by letter dated April 13, 2006, Dominion submitted the revised application changing the cooling design for the proposed Unit 3 and increase in the power level for proposed Units 3 and 4. This supplement 1 to NUREG-1835 documents the staff's analysis of the safety aspects of the cooling design and increase in power level changes.

The North Anna ESP site is located approximately 40 miles north-northwest of Richmond, Virginia, and is adjacent to two existing nuclear power reactors operated by Virginia Electric and Power Company, which, like Dominion Nuclear North Anna, LLC, is a subsidiary of Dominion Resources, Inc. In its application, Dominion seeks an ESP that could support a future application to construct and operate one or more additional nuclear power reactors at the ESP site, with a total nuclear generating capacity of up to 9000 megawatts (thermal). This SER presents the results of the staffs review of information submitted in conjunction with the ESP application. The staff has identified, in Appendix A to this SER, certain site-related items that will need to be addressed at the combined license or construction permit stage, should an applicant desire to construct one or more new nuclear reactors on the North Anna ESP site. Appendix A to this SER also identifies the proposed permit conditions that the staff recommends the Commission impose, should an ESP be issued to the applicant.

12. KEY WORDS/DESCRIPTORS (List words or phrases that will assist researchers in locating the report.)

Early Site Permit (ESP)
Combined License (COL)
Permit Conditions
COL Action Items
Site Characteristics
Bounding Parameters
North Anna ESP Site

13. AVAILABILITY STATEMENT

unlimited

14. SECURITY CLASSIFICATION

(This Page)

unclassified

(This Report)

unclassified

15. NUMBER OF PAGES

16. PRICE